10 Massachusetts MCAS Grade 7 Math Practice Tests

The Ultimate Test Prep Collection with Answer Explanations

Dr. A. Nazari

10 Practice Tests

Grade 7 Math — Engineered for Mastery

Welcome to **The Architect's Workshop.**

Ten practice tests. Ten floors of a tower you're about to build. Each test lays another level of understanding — from foundation to capstone.

Foundation (Tests 1–3): Learn the blueprints

Framework (Tests 4–6): Build structural strength

Finishing (Tests 7–9): Refine under pressure

Capstone (Test 10): The final inspection

Precision. Practice. Perfection.

> **"** Every great structure
> is built one floor at a time.
> Ten levels of practice makes
> your math rock-solid. **"**

The Architectural Plan

A 4-phase construction plan for complete mastery

Phase I: Foundation (Tests 1–3)

Untimed. Explore the test format and question types. Read every answer explanation after each test. Goal: understand the blueprint before you build.

Phase II: Framework (Tests 4–6)

Add a timer (70 minutes). Focus on the topics that gave you trouble in Phase I. Practice showing complete work and labeling units. Goal: build structural strength.

Phase III: Finishing (Tests 7–9)

Full timed conditions (60 minutes). Simulate the real exam. Review only the questions you missed — don't re-study what you already know. Goal: refine under real pressure.

Phase IV: Capstone (Test 10)

Your final inspection. Full exam conditions. This is the capstone — compare with Test 1 and measure your entire growth arc. Goal: prove your mastery.

Blueprint Specifications

Multiple Choice — select the single best answer

Multi-Select — choose ALL that apply

Short Answer — show your process

Open Response — explain and justify

📐 Engineering Principles 📐

📐 The D.R.A.F.T. Method

D **Define** What exactly is the question asking? Write it in your own words.

R **Retrieve** Pull out given information. List numbers, units, and relationships.

A **Assemble** Choose the right formula or strategy. Set up the equation or proportion.

F **Figure** Compute step by step. Show every operation. Label units.

T **Test** Does the answer make sense? Re-read the question. Verify the units.

📐 Seven Precision Rules

1. **Read twice, solve once.** The first read gives context; the second reveals what to compute.

3. **Estimate before calculating.** A quick mental approximation catches major errors.

5. **Track your signs.** Rational number operations are the #1 error source in Grade 7.

7. **Never submit blanks.** On open response, even a partial setup can earn credit.

🕐 Timing Blueprint

Tests 1–3: **Untimed** (learn the blueprints) › Tests 4–6: **70 min** (build speed) › Tests 7–10: **60 min** (exam conditions)

Every great structure starts with a solid plan. Study the blueprints, learn from each test, and build something extraordinary.

Get Online

Find more at
ViewMath.com/MA-Grade7

The Drafting Toolkit

Prepare your station before every construction session

Required Equipment

	Precision Pencils	Two or more #2 pencils, kept sharp. A dull tool produces dull work.
	Quality Eraser	Clean, smudge-free corrections. Architects revise — so will you.
	Scratch Paper	Your drafting workspace. All calculations happen here first.
	Ruler / Protractor	Precision tools for scale drawings, angle measures, and geometry.
	Timer	Begin using from Phase II (Test 4) onward.
	Quiet Workspace	A clean, well-lit station free from distractions.

Approved Materials

- ✔ Pencil and eraser
- ✔ Scratch paper (provided on test day)
- ✔ Ruler or protractor (if specified)
- ✔ Reference sheet (if provided)

Restricted Materials

- ✘ Calculator (unless specified)
- ✘ Electronic devices
- ✘ Notes or reference materials
- ✘ Communication with others

♥ *Project Manager's Guide (for Parents & Guardians)*

Ten practice tests provide the most comprehensive preparation available. The 4-phase structure (Foundation → Framework → Finishing → Capstone) ensures skills build progressively and durably.

Keys to a successful build:

- *Space tests 2–3 days apart — never more than one per day*
- *Phases I–II: review answers together and discuss strategies*
- *Phases III–IV: let them work independently, then debrief results*
- *Compare Test 1 with Test 10 to celebrate the full construction arc*

Find more at
ViewMath.com/MA-Grade7

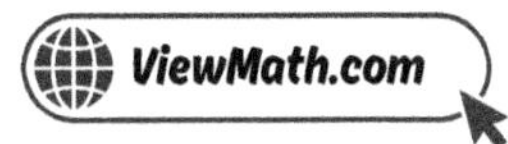

Math Reference Sheet

You may use this page during your practice tests!

Symbol	Name	What It Means	
$(\ )$	Parentheses	Do this part first.	$(3+4) \times 2 = 14$
10^3	Exponent	Multiply the base by itself that many times. $10^3 = 1{,}000$	
$\dfrac{a}{b}$	Fraction	a parts out of b equal parts; also means $a \div b$.	
$\dfrac{7}{3}$	Improper Fraction	Numerator $\geq$ denominator.	$\frac{7}{3} = 2\frac{1}{3}$
0.45	Decimal	A number with a decimal point.	$0.45 = \frac{45}{100}$
$> < =$	Comparison	Greater than, less than, equal to.	$0.5 > 0.35$
$(3,5)$	Ordered Pair	A point on the coordinate plane: (x, y).	

Key Formulas

- **Volume of a rectangular prism:**

  $V = l \times w \times h$

- **Order of operations:**

  Parentheses $\rightarrow$ Exponents $\rightarrow$ Multiply/Divide $\rightarrow$ Add/Subtract

- **Powers of 10:**

  $10^1 = 10 \quad 10^2 = 100$

  $10^3 = 1{,}000 \quad 10^4 = 10{,}000$

- **Fraction as division:**

  $\frac{a}{b} = a \div b$

Place Value Chart

Millions	1,000,000	**Decimals**	
Hundred-Thousands	100,000	Tenths	0.1
Ten-Thousands	10,000	Hundredths	0.01
Thousands	1,000	Thousandths	0.001
Hundreds	100		
Tens	10		
Ones	1		

Each place is $10\times$ the place to its right, and $\frac{1}{10}$ of the place to its left.

Key Math Vocabulary

- **Sum** — the result of addition
- **Difference** — the result of subtraction
- **Product** — the result of multiplication
- **Quotient** — the result of division
- **Remainder** — what's left over after dividing
- **Factor** — a number you multiply
- **Expression** — numbers and operations without $=$
- **Equation** — a math sentence with $=$
- **Numerator** — the top number of a fraction
- **Denominator** — the bottom number of a fraction
- **Mixed number** — a whole number + a fraction
- **Equivalent fractions** — fractions with equal value
- **Decimal** — a number written with a decimal point
- **Volume** — the space inside a 3D shape
- **Coordinate plane** — a grid with x and y axes
- **Ordered pair** — (x, y) location on the plane

Word Problem Clue Words

- **Add** $(+)$: total, altogether, combined, sum, increase, more than
- **Subtract** $(-)$: difference, how many more, fewer, remain, decrease, left
- **Multiply** $(\times)$: each, every, groups of, times, product, per, of (with fractions)
- **Divide** $(\div)$: share equally, split, each group, how many groups, quotient, per

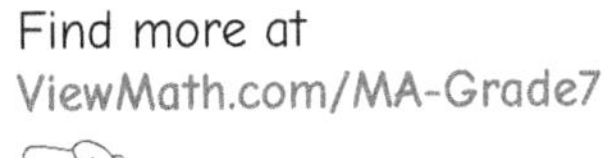
Find more at
ViewMath.com/MA-Grade7

⬛ Multiplication Table ⬛

You may use this table during your practice tests!

×	1	2	3	4	5	6	7	8	9	10	11
1	1	2	3	4	5	6	7	8	9	10	11
2	2	4	6	8	10	12	14	16	18	20	22
3	3	6	9	12	15	18	21	24	27	30	33
4	4	8	12	16	20	24	28	32	36	40	44
5	5	10	15	20	25	30	35	40	45	50	55
6	6	12	18	24	30	36	42	48	54	60	66
7	7	14	21	28	35	42	49	56	63	70	77
8	8	16	24	32	40	48	56	64	72	80	88
9	9	18	27	36	45	54	63	72	81	90	99
10	10	20	30	40	50	60	70	80	90	100	110
11	11	22	33	44	55	66	77	88	99	110	121

💡 How to Use This Table

To find **4 × 7**:

1. Find **4** in the left column (blue).
2. Find **7** in the top row (blue).
3. Follow the row and column until they meet: the answer is **28**!

ℹ Tip: You can also use this table for division! If you know 28 ÷ 4 = ?, find 28 in the 4's row. The column header gives you the answer: **7**!

🏢 Construction Log 🏢

Architect: _______________________________   Project Start: _______________

Phase I: Foundation — Floors 1–3

📐 **Floor 1**   Date: __________   Score: ______ / ______   %: ______
📐 **Floor 2**   Date: __________   Score: ______ / ______   %: ______
📐 **Floor 3**   Date: __________   Score: ______ / ______   %: ______

Phase II: Framework — Floors 4–6

📐 **Floor 4**   Date: __________   Score: ______ / ______   %: ______
📐 **Floor 5**   Date: __________   Score: ______ / ______   %: ______
📐 **Floor 6**   Date: __________   Score: ______ / ______   %: ______

Phase III: Finishing — Floors 7–9

📐 **Floor 7**   Date: __________   Score: ______ / ______   %: ______
📐 **Floor 8**   Date: __________   Score: ______ / ______   %: ______
📐 **Floor 9**   Date: __________   Score: ______ / ______   %: ______

Phase IV: Capstone — Floor 10

📐 **Floor 10**   Date: __________   Score: ______ / ______   %: ______

Certification Level: Apprentice (0–39%)   Technician (40–59%)   Engineer (60–79%)   **Master Architect (80–100%)**

Record your certification level after each test!

Under Construction

Construction Complete

★ Table of Contents ★

Here's what we'll explore together!

 *Let's learn and have fun!*

Practice Test 1

☑ *30 Questions*

✏ Before You Start ✏

- ✔ **Read each question carefully** before choosing your answer.
- ✔ **Show your work** on scratch paper when you need to.
- ✔ **Skip hard questions** and come back to them later.
- ✔ **Check your answers** when you're done.
- ✔ **Take your time** — there's no rush!

⭐ *You've Got This!* ⭐

Do your best and show what you know!

1. The table shows a proportional relationship. Find k and the missing value.

x	y
3	7.5
?	15
10	25

Your Answer

2. Two proportional lines are graphed below. Which statement is correct?

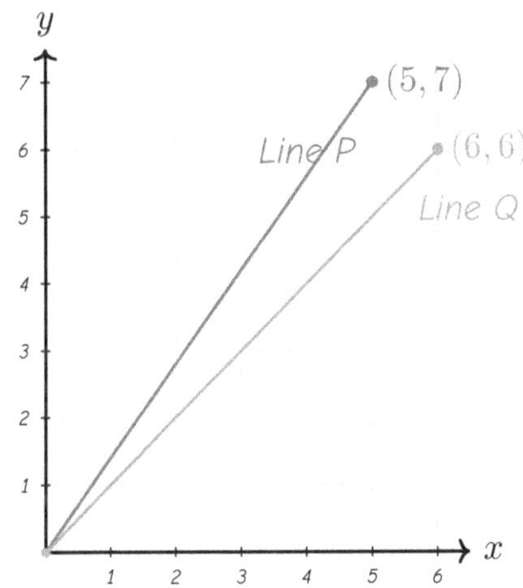

(A) Line Q has the greater unit rate because it is longer

(B) Line P has the greater unit rate because it is steeper

(C) Both lines have the same unit rate

(D) Line Q has the greater unit rate because $6 > 5$

3. The graph below compares the rates of two workers packaging boxes. Which worker is faster, and by how many boxes per hour?

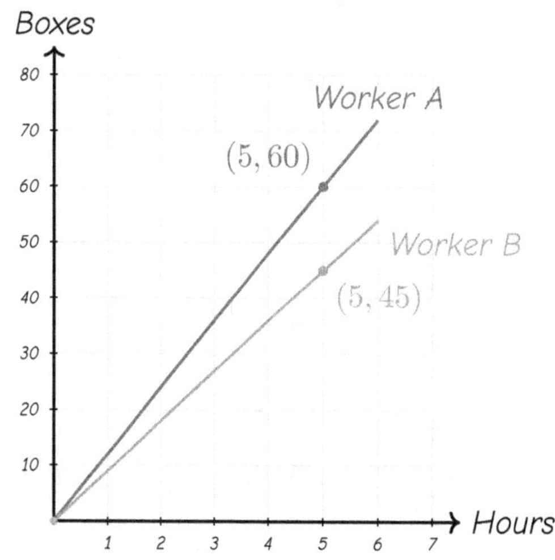

(A) Worker A, by 3 boxes/hr

(B) Worker B, by 3 boxes/hr

(C) Worker A, by 15 boxes/hr

(D) They work at the same rate

4. A store has 300 items in stock. If 60% are on sale, how many items are on sale?

(A) 120

(B) 150

(C) 180

(D) 200

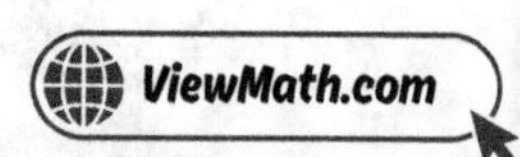

5. The diagram shows how a salesperson's earnings are calculated. What are the total earnings?

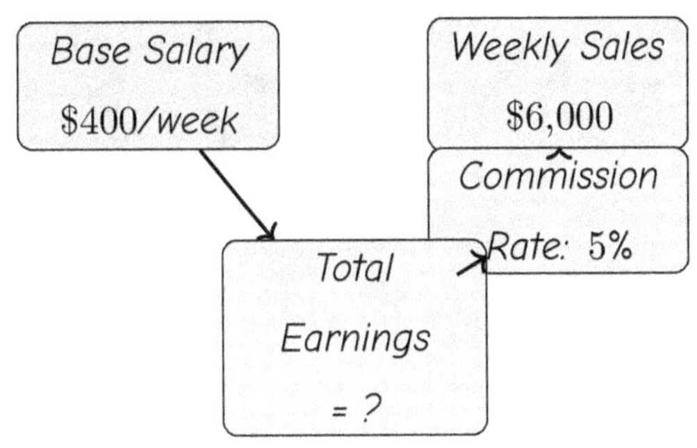

(A) $430

(B) $700

(C) $6,400

(D) $6,700

6. A jogger estimated running 5 miles but actually ran 4.8 miles. What is the percent error (rounded to the nearest tenth)?

Your Answer:

7. The number line below shows two points. Which statement is correct?

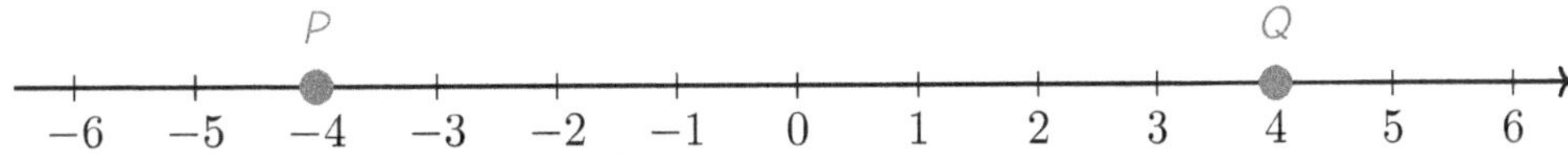

(A) P and Q are the same number

(B) P and Q are opposites

(C) $|P| > |Q|$

(D) $P > Q$

8. What is $0 + (-7)$?

(A) 7

(B) 0

(C) −7

(D) −14

Find more at
ViewMath.com/MA-Grade7

9. Evaluate $\frac{x^2-1}{4}$ when $x = 5$.

Your Answer:

10. Simplify $\frac{1}{2}x + \frac{3}{4}x - 2$.

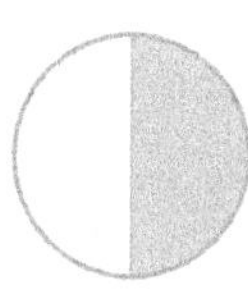

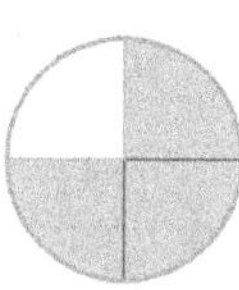

Your Answer:

11. The diagram shows a triangle with a perimeter of 39 cm. What is the value of x?

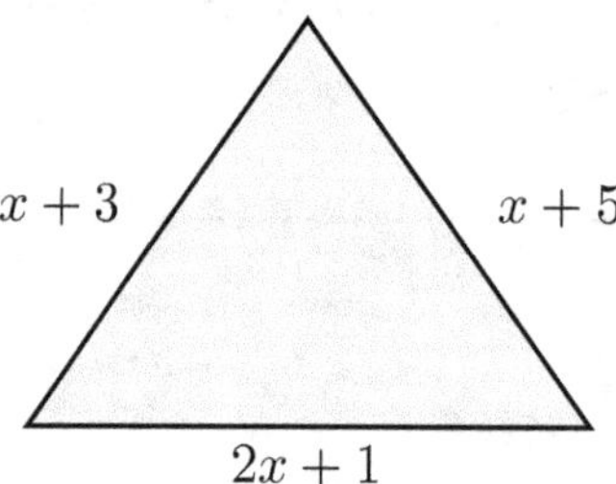

(A) $x = 7.5$ (B) $x = 10$

(C) $x = 6$ (D) $x = 8$

Find more at
ViewMath.com/MA-Grade7

ViewMath.com

12. Four friends go to a museum. Each person pays for a ticket and a $3 audio guide. The total for all four people is $68. The equation $4(t+3) = 68$ represents this situation, where t is the ticket price. Use the model below to find t.

4 people $\left.\begin{array}{|c|c|} \hline t & \$3 \\ \hline t & \$3 \\ \hline t & \$3 \\ \hline t & \$3 \\ \hline \end{array}\right\} \$68$

Your Answer

13. A store marks down a jacket by $\dfrac{1}{4}$ of its original price p, then subtracts an additional $8. The sale price is $37. Find the original price.

Your Answer

14. Solve $2x + 3 > 11$.

(A) $x > 4$

(B) $x > 7$

(C) $x < 4$

(D) $x > 14$

15. What inequality is shown by a closed circle at -6 with shading to the right?

Your Answer

16. *A drawing uses $1\ cm = 3\ m$. It is redrawn at $1\ cm = 9\ m$. A room that is $12\ cm$ on the original becomes how long?*

   (A) 4 cm   (B) 12 cm

   (C) 36 cm   (D) 108 cm

17. *Two sides of a triangle are 5 cm and 5 cm with an included angle of 60°. What type of triangle is formed?*

Your Answer:

18. *What condition (SSS, SAS, ASA, AAS, or AAA) gives infinitely many triangles?*

Your Answer:

19. *You cut a rectangular prism with a cut that passes through exactly three faces. What is the cross-section shape?*

   (A) Rectangle   (B) Triangle

   (C) Pentagon   (D) Hexagon

20. *A circle has a diameter of 14 cm. What is its area? Use $\pi \approx \frac{22}{7}$.*

   (A) $44\ cm^2$ (B) $154\ cm^2$

   (C) $308\ cm^2$ (D) $616\ cm^2$

Find more at
ViewMath.com/MA-Grade7

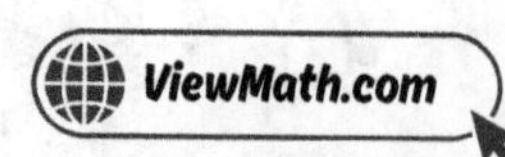

21. A figure is made of a rectangle 12 cm by 5 cm with a right triangle removed. The triangle has legs 4 cm and 5 cm. What is the area of the remaining figure?

(A) 40 cm^2

(B) 50 cm^2

(C) 60 cm^2

(D) 70 cm^2

22. A triangular prism has a triangular base with base 6 cm and height 4 cm, and the prism is 10 cm long. The three rectangular faces have widths 6 cm, 5 cm, and 5 cm. What is the total surface area?

(A) 184 cm^2

(B) 160 cm^2

(C) 124 cm^2

(D) 120 cm^2

23. A rectangular fish tank is 50 cm long, 30 cm wide, and 40 cm tall. What is the volume?

(A) 120 cm^3

(B) 6,000 cm^3

(C) 60,000 cm^3

(D) 600 cm^3

24. In a school of 600 students, a random sample of 60 found that 15 students walk to school. Predict how many students in the entire school walk to school.

(A) 100

(B) 150

(C) 200

(D) 250

25. Which measure of variability is shown directly on a box plot?

(A) Mean absolute deviation (MAD)

(B) Standard deviation

(C) Interquartile range (IQR)

(D) Mean

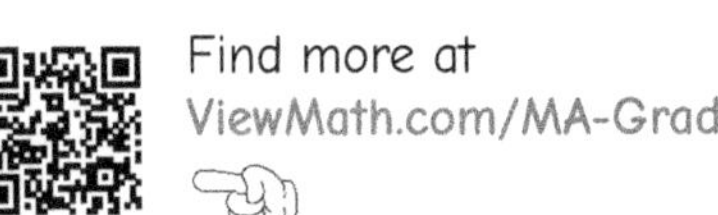

Find more at
ViewMath.com/MA-Grade7

26. The table compares two delivery companies' shipping times (in days).

	Company X	Company Y
Mean	3.5 days	3.2 days
Median	3 days	3 days
Range	4 days	8 days
IQR	1.5 days	4 days

A customer wants the most reliable delivery time. Which company should they choose?

(A) Company Y, because its mean is lower

(B) Company X, because its IQR and range are both smaller

(C) Company Y, because its range includes more possibilities

(D) They are equally reliable

27. A jar has 5 green, 3 yellow, and 2 orange gumballs. What is the probability of NOT picking a green gumball?

(A) 0.5

(B) 0.3

(C) 0.7

(D) 0.2

28. A probability model has outcomes X, Y, and Z. $P(X) = 0.15$ and $P(Y) = 0.45$. What is $P(Z)$?

Your Answer:

29. A spinner has 4 equal sections $(1, 2, 3, 4)$ and a die is rolled. What is the probability that both show a 3?

(A) $\frac{1}{10}$

(B) $\frac{1}{24}$

(C) $\frac{2}{10}$

(D) $\frac{1}{4}$

Find more at
ViewMath.com/MA-Grade7

30. A spinner has a 60% chance of landing on blue. In a simulation of 80 spins, how many times would you expect blue to appear?

Your Answer:

 End of Practice Test 1 

Great job finishing the test!

 My Score

I got _____________ out of 30 questions right.

*Check your answers in the **Answer Key** at the back of the book.*

 *Review any questions you missed. That's how we learn!*

Check Your Score Online!

Visit **ViewMath Academy** to enter your answers and see which topics you need to review. You can also explore lessons, take quizzes, track your scores, and save your progress!

viewmath.com/score/7.1.MA.16

Or go to viewmath.com/score and enter code: 7.1.MA.16

2

Practice Test 2

📋 *30 Questions*

✏️ Before You Start ✏️

- ✓ **Read each question carefully** before choosing your answer.
- ✓ **Show your work** on scratch paper when you need to.
- ✓ **Skip hard questions** and come back to them later.
- ✓ **Check your answers** when you're done.
- ✓ **Take your time** — there's no rush!

⭐ You've Got This! ⭐

Do your best and show what you know!

1. The table below shows a proportional relationship between gallons of gas and miles driven. What is k and what does it mean?

Gallons (x)	Miles (y)
2	54
5	135
8	216

(A) $k = 2$; uses 2 gallons per trip

(B) $k = 27$; drives 27 miles per gallon

(C) $k = 54$; drives 54 miles on 2 gallons

(D) $k = 135$; drives 135 miles on 5 gallons

2. The graph below shows the number of birdhouses a carpenter can build over time. What is the unit rate, and how many birdhouses can be built in 10 hours?

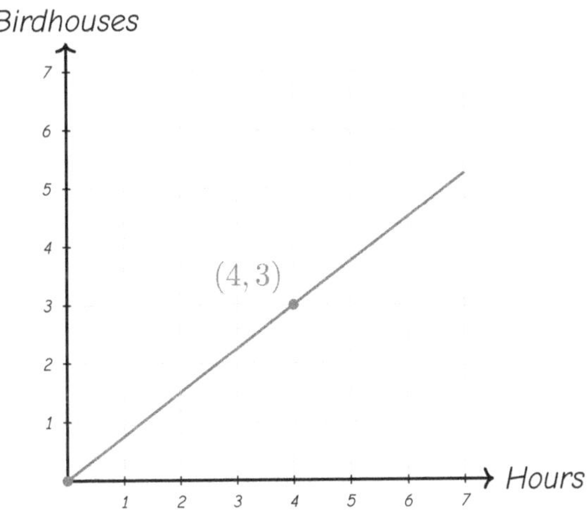

Your Answer

3. The table shows how many batches of muffins a bakery can make and the total flour used. How many cups of flour are needed for 15 batches?

Batches	Flour (cups)
3	$7\frac{1}{2}$
6	15
9	$22\frac{1}{2}$
15	?

Your Answer:

4. A class has 32 students. If 75% passed the test, how many students passed?

(A) 20

(B) 24

(C) 25

(D) 28

5. A bill is $86 and you leave a $17.20 tip. What percent tip is that?

(A) 15%

(B) 17%

(C) 18%

(D) 20%

6. Two students estimated the number of beans in a jar. The actual count was 200. Student A guessed 180; Student B guessed 220. Who had the greater percent error?

(A) Student A

(B) Student B

(C) Same percent error

(D) Cannot be determined

Find more at
ViewMath.com/MA-Grade7

7. What is $|-9|$?

   (A) -9 (B) 0

   (C) $\frac{1}{9}$ (D) 9

8. Find the sum: $5 + (-11) + 3 + (-2)$.

Your Answer:

9. What is the value of $\frac{x}{2} - 4$ when $x = 10$?

   (A) 3 (B) -1

   (C) 1 (D) 9

10. Simplify $-2a + 5b + 7a - 3b$.

Your Answer:

11. Solve $8 = \frac{x}{3} + 2$.

   (A) $x = 30$ (B) $x = 18$

   (C) $x = 2$ (D) $x = 6$

12. Solve $2(x + 7) = -6$.

Your Answer:

13. A number line shows a point at x. The equation $\frac{x}{4} + 1.5 = 5$ tells you how to find x. Solve and identify where x is on the number line.

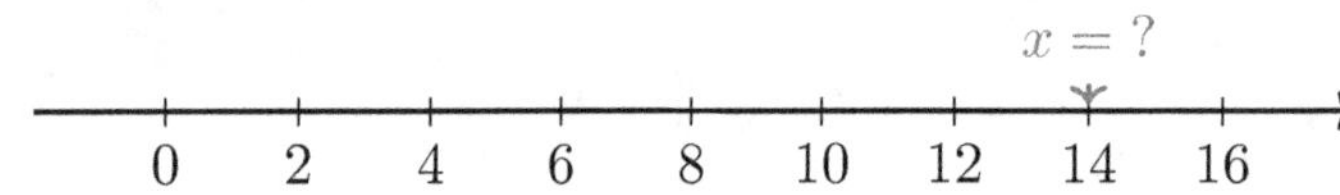

Your Answer

14. Solve $\frac{x}{2} + 7 \le 12$.

Your Answer

15. Which inequality is shown by a closed circle at 7 with shading to the left?

(A) $x > 7$

(B) $x < 7$

(C) $x \ge 7$

(D) $x \le 7$

16. A map uses 1 cm = 20 km. You redraw it at 1 cm = 10 km. What happens to the drawing?

(A) Every length is halved

(B) Every length stays the same

(C) Every length is doubled

(D) Every length is multiplied by 10

17. You know all three angles of a triangle: 50°, 60°, and 70°. How many triangles can be drawn?

(A) None

(B) Exactly one

(C) Exactly three

(D) More than one (infinitely many)

18. Can a triangle be formed with sides 7, 10, and 15?

(A) No, the triangle inequality fails

(B) Yes, exactly one triangle

(C) Yes, more than one triangle

(D) Yes, but only a right triangle

19. A rectangular prism has a base of 6 cm by 4 cm and a height of 10 cm. It is sliced parallel to the base. What are the dimensions of the cross-section?

Your Answer

20. A circle has an area of 78.5 cm^2. Using $\pi \approx 3.14$, what is the radius?

(A) 4 cm

(B) 5 cm

(C) 10 cm

(D) 25 cm

21. A swimming pool is shaped like a rectangle 25 m by 10 m with a semicircle of diameter 10 m at one end. What is the total area? Use $\pi \approx 3.14$.

(A) 250 m^2

(B) 289.25 m^2

(C) 328.5 m^2

(D) 367.75 m^2

22. A triangular prism has a right-triangle base with legs 3 cm and 4 cm. The prism is 12 cm long. The hypotenuse of the triangle is 5 cm. What is the total surface area?

Your Answer

Find more at
ViewMath.com/MA-Grade7

23. A triangular prism has a base that is a triangle with base 8 cm and height 3 cm. The prism is 10 cm long. What is the volume?

(A) $24\ cm^3$

(B) $80\ cm^3$

(C) $120\ cm^3$

(D) $240\ cm^3$

24. A factory produces 10,000 batteries per day. A sample of 200 batteries is tested, and 8 are defective. What percent of the sample is defective?

Your Answer:

25. The dot plots below show the number of push-ups completed by students in two gym classes.

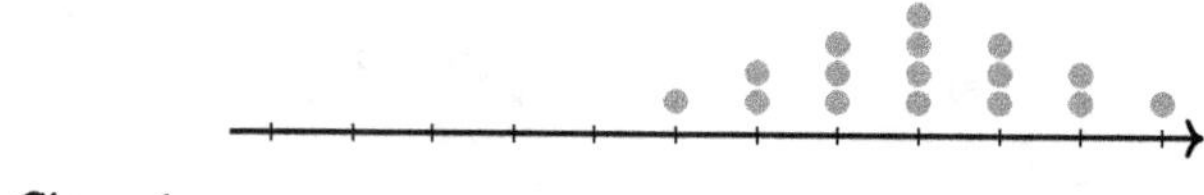
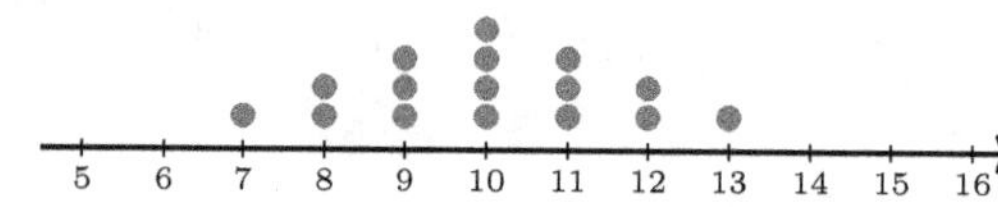

Which class has a higher center, and by approximately how many push-ups?

(A) Class 1, by about 3

(B) Class 2, by about 3

(C) Class 2, by about 6

(D) They have the same center

Find more at
ViewMath.com/MA-Grade7

26. The dot plots show daily high temperatures (°F) for two cities over 10 days.

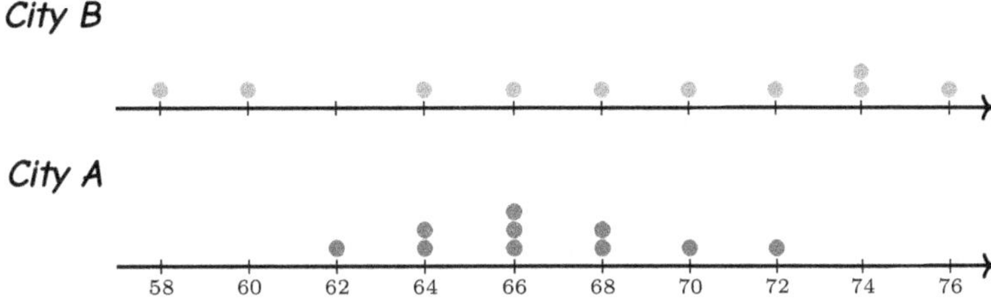

Find the mean and range for each city. Which city has more consistent temperatures?

Your Answer

27. What does a probability of 0 mean?

(A) The event is certain to happen

(B) The event is very likely to happen

(C) The event is impossible

(D) The event has a 50% chance of happening

28. A student creates a probability model for a spinner. $P(red) = 0.25$, $P(blue) = 0.25$, $P(green) = 0.25$, $P(yellow) = 0.25$. After 200 spins, the results are: red $= 55$, blue $= 48$, green $= 52$, yellow $= 45$. Is the model a good fit?

(A) No, the model is completely wrong

(B) Yes, the observed frequencies are reasonably close to 50 each

(C) No, because the results are not exactly 50 each

(D) Yes, because 200 is divisible by 4

Find more at
ViewMath.com/MA-Grade7

29. The tree diagram shows the outcomes for flipping two coins. What is the probability of getting at least one tail?

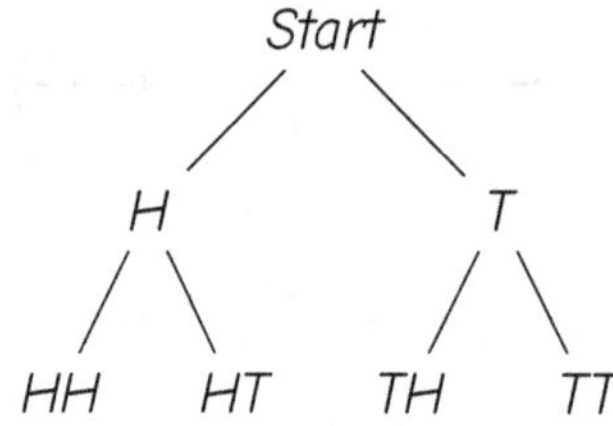

(A) $\frac{1}{4}$

(B) $\frac{1}{2}$

(C) $\frac{3}{4}$

(D) $\frac{2}{4}$

30. *The bar graph shows the results of simulating a spinner* 100 *times. The spinner is supposed to land on each color with equal probability. Based on the simulation, does the spinner appear fair?*

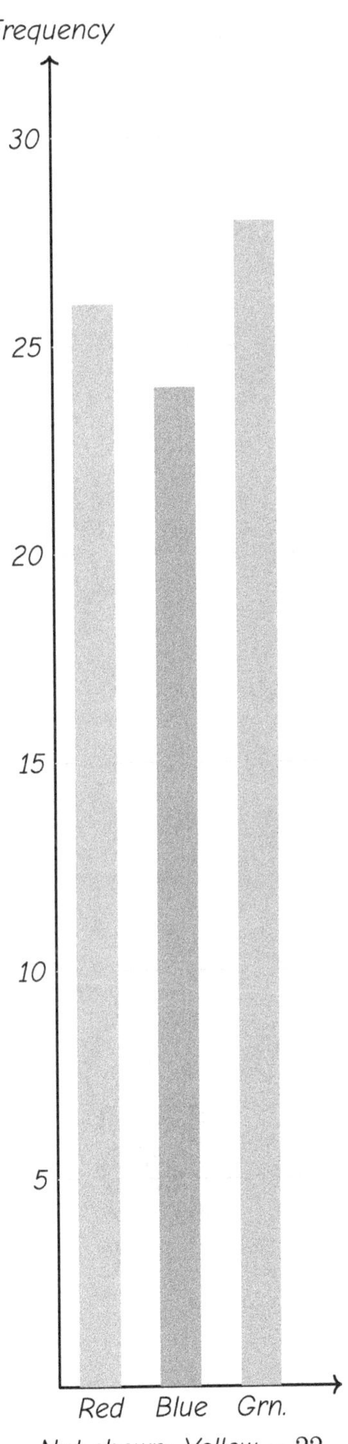

(A) *No, because the bars are different heights*

(B) *Yes, the frequencies* $(26, 24, 28, 22)$ *are all roughly close to* 25

(C) *Green is clearly the most likely color*

(D) *Cannot be determined from the graph*

 # End of Practice Test 2 

Great job finishing the test!

☑ My Score

I got ___________ out of 30 questions right.

*Check your answers in the **Answer Key** at the back of the book.*

💡 Review any questions you missed. That's how we learn!

📊 Check Your Score Online!

Visit **ViewMath Academy** to enter your answers and see which topics you need to review. You can also explore lessons, take quizzes, track your scores, and save your progress!

viewmath.com/score/7.1.MA.17

Or go to **viewmath.com/score** and enter code: 7.1.MA.17

Practice Test 3

30 Questions

✏️ Before You Start ✏️

- ✓ **Read each question carefully** before choosing your answer.
- ✓ **Show your work** on scratch paper when you need to.
- ✓ **Skip hard questions** and come back to them later.
- ✓ **Check your answers** when you're done.
- ✓ **Take your time** — there's no rush!

⭐ You've Got This! ⭐

Do your best and show what you know!

1. A proportional table has $k = \frac{3}{4}$. Which pair could be in the table?

(A) $(8, 10)$

(B) $(8, 6)$

(C) $(12, 8)$

(D) $(4, 2)$

2. Two proportional lines are graphed. Line A is steeper than Line B. What does this tell you?

(A) Line A has a smaller unit rate

(B) Line A has a greater unit rate

(C) Line B covers more distance

(D) Both lines have the same rate

3. Store A sells 3 notebooks for $7.50. Store B sells 5 notebooks for $11.25. Which store offers the better deal?

(A) Store A at $2.50 each

(B) Store B at $2.25 each

(C) Store A at $2.25 each

(D) They cost the same

4. 63 is 90% of what number?

(A) 56.7

(B) 67

(C) 70

(D) 72

5. Estimate a 15% tip on a $78 bill using the 10% method.

(A) About $7.80

(B) About $10.40

(C) About $11.70

(D) About $15.60

6. A student measured a table as 4.5 feet long. The actual length is 5 feet. What is the percent error?

(A) 0.5%

(B) 5%

(C) 10%

(D) 11.1%

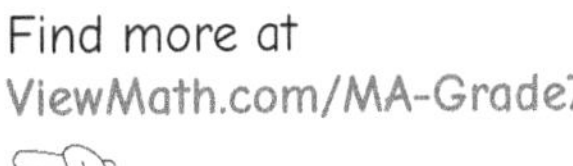
Find more at
ViewMath.com/MA-Grade7

ViewMath.com

7. For which value of n is $|n| = 10$?

(A) Only $n = 10$

(B) Only $n = -10$

(C) $n = 10$ or $n = -10$

(D) $n = 0$

8. What is $15 + (-15)$?

(A) 30

(B) -30

(C) 15

(D) 0

9. What is the value of $x^2 + 3x$ when $x = -2$?

(A) 10

(B) -2

(C) -10

(D) 2

10. Simplify $-6y + 4y - y$.

(A) $-3y$

(B) $-y$

(C) $3y$

(D) $-11y$

11. Solve $9 + 3x = 30$.

(A) $x = 13$

(B) $x = 3$

(C) $x = 7$

(D) $x = 10$

12. Solve $9(x - 3) = 45$.

Your Answer:

13. *Solve* $-0.4y + 2 = 6$.

(A) $y = 10$ (B) $y = -10$

(C) $y = 20$ (D) $y = -20$

14. *Solve* $-5n + 10 \leq -15$.

Your Answer:

15. *Solve* $-3x + 12 \geq 0$ *and describe the graph.*

(A) *Closed circle at 4, shade right*   (B) *Open circle at 4, shade left*

(C) *Closed circle at 4, shade left*   (D) *Closed circle at −4, shade right*

16. *A map uses* $1 \text{ cm} = 60 \text{ km}$. *You redraw it at* $1 \text{ cm} = 20 \text{ km}$. *A triangle on the original has an area of* 4 cm^2. *What is its area on the new map?*

(A) 12 cm^2 (B) 16 cm^2

(C) 36 cm^2 (D) 48 cm^2

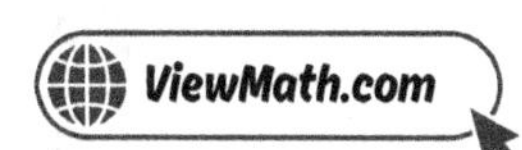

17. The diagram shows a triangle with the given measurements. How many different triangles can be drawn with these exact conditions?

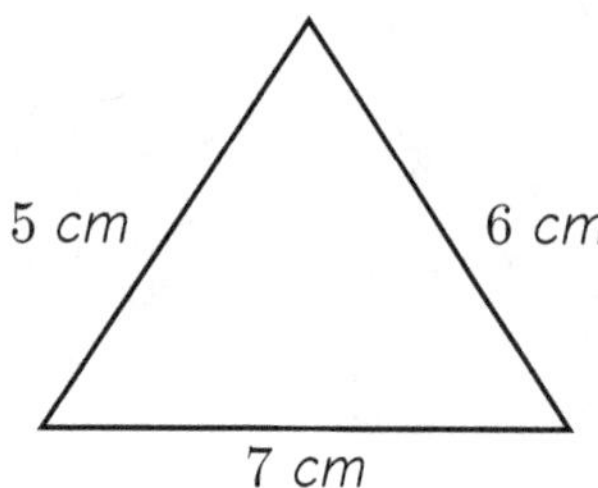

(A) None

(B) Exactly one

(C) Exactly two

(D) Infinitely many

18. A triangle has angles 90°, 45°, and 45°. Which statement is true?

(A) Exactly one such triangle exists

(B) Many triangles with these angles exist, in different sizes

(C) No such triangle exists because two angles are equal

(D) No such triangle exists because the angles don't sum to 180°

19. A triangular prism is sliced with a vertical cut perpendicular to the triangular bases. What shape is the cross-section?

(A) Triangle

(B) Rectangle

(C) Circle

(D) Trapezoid

20. A circular rug has a diameter of 8 feet. What is the area of the rug? Use $\pi \approx 3.14$

(A) 25.12 ft^2

(B) 50.24 ft^2

(C) 200.96 ft^2

(D) 64 ft^2

Find more at
ViewMath.com/MA-Grade7

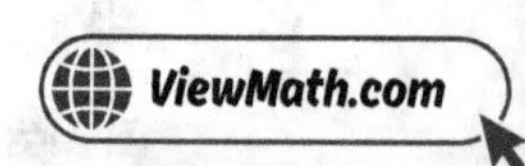

21. A figure is made of a right triangle with legs 6 cm and 8 cm, and a rectangle 8 cm by 3 cm. What is the total area?

(A) 24 cm^2

(B) 48 cm^2

(C) 36 cm^2

(D) 72 cm^2

22. The net shown below folds into a rectangular prism. Find the total surface area.

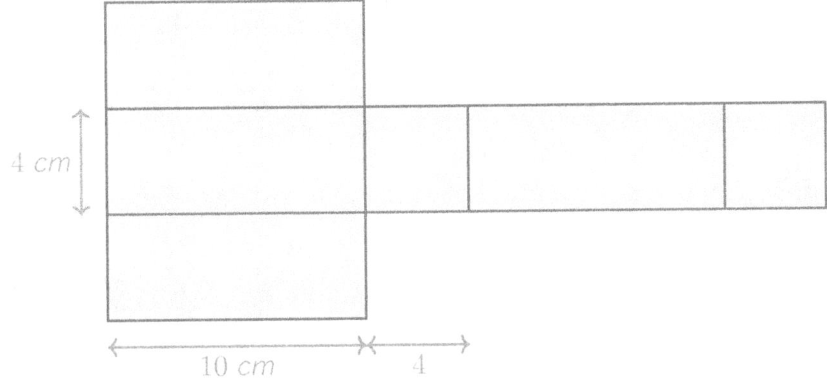

Your Answer

23. A trapezoidal prism has a trapezoidal base with parallel sides 4 cm and 6 cm, height 3 cm, and the prism length is 10 cm. What is its volume?

(A) 60 cm^3

(B) 120 cm^3

(C) 150 cm^3

(D) 180 cm^3

24. Which of the following increases the reliability of conclusions drawn from a sample?

(A) Using a smaller sample

(B) Using a larger random sample

(C) Surveying only people who agree with you

(D) Using a non-random sample

Find more at
ViewMath.com/MA-Grade7

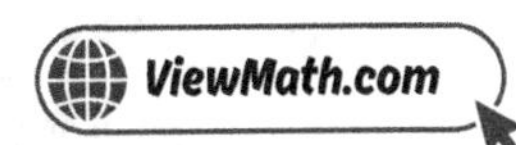

25. *School A: median score 88, MAD 4. School B: median score 82, MAD 4. The difference in medians expressed in MADs is:*

(A) 1 MAD

(B) 1.5 MADs

(C) 2 MADs

(D) 4 MADs

26. *Group X: mean 45, MAD 6. Group Y: mean 57, MAD 6. Express the difference in means as a multiple of MAD.*

Your Answer:

27. *Jasmine says the probability of picking a vowel from the letters of the word MATH is $\frac{2}{4}$. Is she correct?*

(A) Yes, because there are 2 vowels in MATH

(B) No, there is only 1 vowel in MATH, so the probability is $\frac{1}{4}$

(C) No, the probability is $\frac{3}{4}$

(D) Yes, because half of all letters are vowels

28. *A bag contains 4 red marbles and 1 blue marble. Which type of probability model does this represent?*

(A) Uniform

(B) Non-uniform

(C) Both uniform and non-uniform

(D) Neither

29. *Two dice are rolled. What is the probability of getting a sum of 2?*

(A) $\frac{1}{6}$

(B) $\frac{1}{12}$

(C) $\frac{1}{36}$

(D) $\frac{2}{36}$

Find more at
ViewMath.com/MA-Grade7

30. You simulate flipping a coin 3 times and counting how many times all 3 are heads. In 40 simulations, all 3 heads occurs 6 times. What is the experimental probability?

(A) $\frac{6}{40} = 0.15$

(B) $\frac{1}{8} = 0.125$

(C) $\frac{3}{40} = 0.075$

(D) $\frac{6}{120} = 0.05$

Find more at
ViewMath.com/MA-Grade7

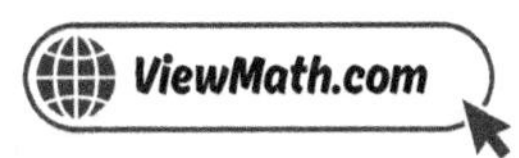

 # End of Practice Test 3 

Great job finishing the test!

 My Score

I got ___________ out of 30 questions right.

Check your answers in the **Answer Key** at the back of the book.

 Review any questions you missed. That's how we learn!

📊 Check Your Score Online!

Visit **ViewMath Academy** to enter your answers and see which topics
you need to review. You can also explore lessons, take quizzes, track
your scores, and save your progress!

viewmath.com/score/7.1.MA.18

Or go to viewmath.com/score and enter code: 7.1.MA.18

Practice Test 4

📋 *30 Questions*

✏️ Before You Start ✏️

- ✓ **Read each question carefully** before choosing your answer.
- ✓ **Show your work** on scratch paper when you need to.
- ✓ **Skip hard questions** and come back to them later.
- ✓ **Check your answers** when you're done.
- ✓ **Take your time** — there's no rush!

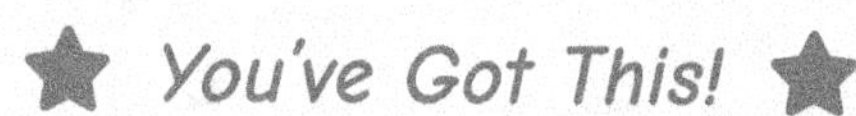

⭐ You've Got This! ⭐

Do your best and show what you know!

1. What is the constant of proportionality (k) for the table below?

x	y
3	12
5	20
8	32

(A) 3

(B) 4

(C) 8

(D) 12

2. A proportional graph shows total cost versus hours of tutoring. The point $(3, 67.50)$ is on the graph. How much does 5 hours of tutoring cost?

Your Answer:

3. A factory produces 168 items in 7 hours. Another factory produces 200 items in 8 hours. Which factory is more productive?

(A) Factory 1, at 24 items/hr

(B) Factory 2, at 25 items/hr

(C) They produce the same rate

(D) Factory 1, at 25 items/hr

4. A jar has 250 candies. If 44% are chocolate, how many chocolate candies are there?

Your Answer:

5. A restaurant bill is $40. You leave a 15% tip. How much is the tip?

(A) $4.00

(B) $5.00

(C) $6.00

(D) $8.00

Find more at
ViewMath.com/MA-Grade7

6. In the percent error formula, why do we use absolute value?

(A) To make the calculation easier

(B) Because the answer must be positive

(C) Because we only care about the size of the error, not the direction

(D) To convert the answer to a percent

7. Which statement about absolute value is FALSE?

(A) $|-5| = 5$

(B) $|0| = 0$

(C) Absolute value can be negative

(D) $|3| = |-3|$

8. A hiker starts at an elevation of -30 feet (below sea level). She climbs 45 feet. What is her new elevation?

Your Answer

9. Which expression represents "the product of 7 and a number p, decreased by 12"?

(A) $7(p - 12)$

(B) $12 - 7p$

(C) $7p - 12$

(D) $7 + p - 12$

10. Simplify $7x + 3x$.

(A) $10x$

(B) $21x$

(C) $10x^2$

(D) $73x$

11. Solve $\dfrac{w}{5} + 12 = 20$.

(A) $w = 40$

(B) $w = 160$

(C) $w = 8$

(D) $w = 1.6$

Find more at
ViewMath.com/MA-Grade7

12. *Solve* $5(2x - 3) = 35$.

Your Answer:

13. *Use the fraction bar model to solve:* $\dfrac{2}{3}x = 8$. *What is x?*

$$\frac{2}{3}x = 8 \qquad ?$$

$$\frac{1}{3} \qquad \frac{1}{3} \qquad \frac{1}{3}$$

(A) $x = 12$

(B) $x = 16$

(C) $x = 10$

(D) $x = 5\dfrac{1}{3}$

14. *Solve* $8 - 3x \geq 23$.

Your Answer:

15. *Solve* $-2x \leq 10$ *and describe the graph.*

(A) *Closed circle at -5, shade left*

(B) *Closed circle at 5, shade left*

(C) *Closed circle at -5, shade right*

(D) *Open circle at -5, shade right*

16. A triangle on Drawing A (scale 1 cm = 4 m) has the dimensions shown below. Drawing B uses scale 1 cm = 2 m. What is the base of the triangle on Drawing B?

Drawing A

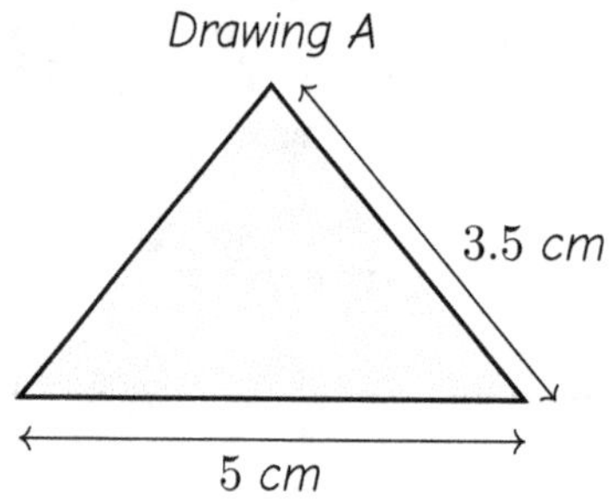

(A) 2.5 cm

(B) 5 cm

(C) 10 cm

(D) 20 cm

17. Marcus wants to draw a triangle with angles 60°, 60°, and 60° and a side of 4 cm. How many different triangles can he draw?

(A) None

(B) Exactly one

(C) Exactly three

(D) Infinitely many

18. Can a triangle have angles 90°, 100°, and −10°?

(A) Yes, it is a right triangle

(B) Yes, because the sum is 180°

(C) No, because an angle cannot be negative

(D) No, because the sum is not 180°

19. Which cross-section is NOT possible from slicing a rectangular prism?

(A) Rectangle

(B) Triangle

(C) Circle

(D) Pentagon

Find more at
ViewMath.com/MA-Grade7

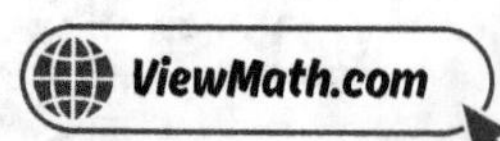

20. What is the area of a circle with a radius of 10 cm? Use $\pi \approx 3.14$.

(A) 31.4 cm^2

(B) 62.8 cm^2

(C) 100 cm^2

(D) 314 cm^2

21. A shape is made of a square with side 8 cm and a semicircle with diameter 8 cm attached to one side. What is the area? Use $\pi \approx 3.14$.

(A) 64 cm^2

(B) 89.12 cm^2

(C) 114.24 cm^2

(D) 139.12 cm^2

22. A rectangular prism has dimensions 7 m by 4 m by 3 m. What is the area of the largest face?

(A) 12 m^2

(B) 21 m^2

(C) 28 m^2

(D) 84 m^2

23. The figure shows unit cubes stacked into an L-shape. Each cube has a side length of 1 cm. What is the total volume?

(A) 8 cm^3

(B) 11 cm^3

(C) 15 cm^3

(D) 33 cm^3

24. Give one reason why a sample might not perfectly represent a population.

Your Answer

25. When comparing two populations visually, you should look at:

(A) Only the center (mean or median)

(B) Only the spread (range or MAD)

(C) Both the center and the spread

(D) Only the largest and smallest values

26. Store A daily sales: mean $500, MAD $40. Store B daily sales: mean $620, MAD $40. The difference in means in MADs is:

(A) 1.5 MADs

(B) 3 MADs

(C) 40 MADs

(D) 120 MADs

27. Marcus says the probability of his basketball team winning tonight is $\frac{3}{4}$. Which statement best describes this?

(A) The team will definitely win

(B) The team is unlikely to win

(C) The team is likely to win, but not certain

(D) The team has an equal chance of winning or losing

28. A coin is weighted so that heads comes up 60% of the time. What is the probability of tails?

(A) 0.60

(B) 0.50

(C) 0.40

(D) 0.30

Find more at
ViewMath.com/MA-Grade7

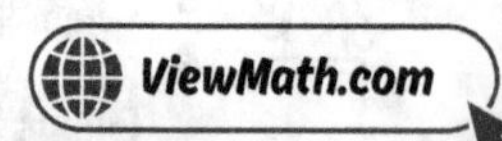

29. Look at the tree diagram for flipping three coins. What is the probability of getting exactly 2 tails? Write your answer as a fraction in simplest form.

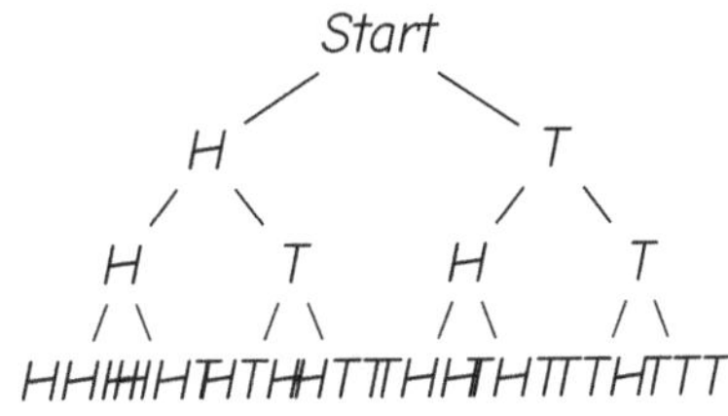

Your Answer:

30. A student wants to simulate the probability that exactly 2 out of 3 students pass a test, where each student has a 50% chance. Which model would work?

(A) Roll one die: odd = pass, even = fail

(B) Flip 3 coins: heads = pass, tails = fail; count exactly 2 heads

(C) Draw 3 marbles from a bag of 2 red and 1 blue

(D) Flip 1 coin 3 times and count all tails

Great job finishing the test!

 **My Score**

I got _________ out of 30 questions right.

Check your answers in the **Answer Key** at the back of the book.

💡 Review any questions you missed. That's how we learn!

📊 **Check Your Score Online!**

Visit **ViewMath Academy** to enter your answers and see which topics you need to review. You can also explore lessons, take quizzes, track your scores, and save your progress!

viewmath.com/score/7.1.MA.19

Or go to viewmath.com/score and enter code: 7.1.MA.19

Practice Test 5

✅ 30 Questions

✏️ Before You Start ✏️

- ✓ **Read each question carefully** before choosing your answer.
- ✓ **Show your work** on scratch paper when you need to.
- ✓ **Skip hard questions** and come back to them later.
- ✓ **Check your answers** when you're done.
- ✓ **Take your time** — there's no rush!

⭐ You've Got This! ⭐

Do your best and show what you know!

1. A factory produces widgets at a constant rate. In $2\frac{1}{2}$ hours it makes 175 widgets. What is k in widgets per hour?

2. A proportional graph passes through $(3, 18)$. What is the unit rate?

(A) 3

(B) 6

(C) 15

(D) 18

3. A 12-oz box of cereal costs $3.60. A 20-oz box costs $5.40. Which is the better value?

(A) 12-oz box at $0.30/oz

(B) 20-oz box at $0.27/oz

(C) 12-oz box at $0.27/oz

(D) They are the same price per ounce

4. The tape diagram below represents a whole quantity. What percent of the total is the shaded portion?

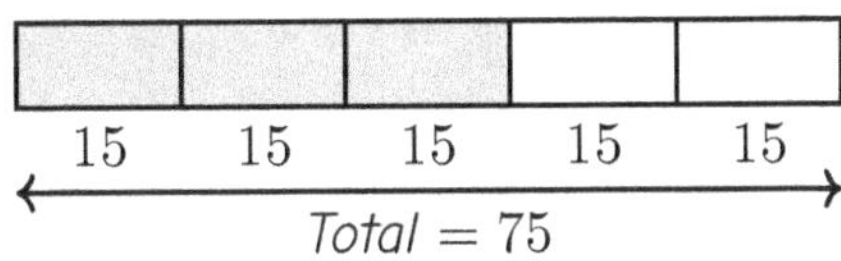

(A) 40%

(B) 45%

(C) 50%

(D) 60%

5. A real estate agent earns 3% commission on a $250,000 home sale. What does she earn?

(A) $750

(B) $2,500

(C) $7,500

(D) $75,000

6. An architect estimated a building would be 45 meters tall. The actual height is 42 meters. What is the percent error (rounded to the nearest tenth)?

Your Answer:

7. Which statement is true?

(A) $-5 > -2$

(B) $-3 > 0$

(C) $-7 < -4$

(D) $0 < -1$

8. What value of n makes $(-9) + n = -2$ true?

(A) -11

(B) -7

(C) 7

(D) 11

9. Which expression means "half the sum of a number k and 10"?

(A) $\frac{k}{2} + 10$

(B) $\frac{k+10}{2}$

(C) $\frac{10k}{2}$

(D) $\frac{k}{2+10}$

10. The perimeter of a triangle is $3x + 2 + 5x - 1 + 2x + 4$. Write the simplified expression for the perimeter.

Your Answer:

11. Solve $2x + 15 = 31$.

Your Answer:

Find more at
ViewMath.com/MA-Grade7

12. Solve $-4(y + 3) = 28$.

  (A) $y = 4$ (B) $y = -10$

  (C) $y = 10$ (D) $y = -4$

13. Solve $-0.3x + 4.2 = 1.8$.

Your Answer:

14. A movie theater requires you to be more than 48 inches tall for a ride. You are 42 inches tall and grow g inches per year. Which inequality finds how many years until you are tall enough?

  (A) $42 + g > 48$ (B) $42g > 48$

  (C) $48 - g > 42$ (D) $42 - g > 48$

15. The graph of an inequality has an open circle at -1 and is shaded to the left. Which inequality matches?

  (A) $x \leq -1$ (B) $x < -1$

  (C) $x \geq -1$ (D) $x > -1$

16. A blueprint uses scale $1 : 80$. You redraw at scale $1 : 20$. A room is 2 cm by 3 cm on the original. What is the area of the room on the new drawing?

Your Answer:

17. A parallelogram has side lengths 5 cm and 8 cm. How many different parallelograms can be drawn?

  (A) None   (B) Exactly one

  (C) Exactly two   (D) More than one

Find more at
ViewMath.com/MA-Grade7

ViewMath.com

18. Can a triangle have angles 50°, 60°, and 80°?

(A) Yes, because the angles are all positive

(B) Yes, because each angle is less than 180°

(C) No, because $50 + 60 + 80 = 190 \neq 180$

(D) No, because the angles are not equal

19. Which 3D figure always has circular cross-sections when cut parallel to its base?

(A) Rectangular prism

(B) Triangular pyramid

(C) Cone

(D) Cube

20. A sprinkler waters a circular area with a radius of 5 meters. How many square meters does it water? Use $\pi \approx 3.14$

(A) $31.4 \ m^2$

(B) $50 \ m^2$

(C) $78.5 \ m^2$

(D) $157 \ m^2$

21. An H-shaped figure is made of three rectangles. The top and bottom bars are each 10 cm by 2 cm. The middle vertical bar is 2 cm by 6 cm. What is the total area?

Your Answer

22. How many faces does a rectangular prism have?

(A) 4

(B) 5

(C) 6

(D) 8

Find more at
ViewMath.com/MA-Grade7

23. A moving box is 2 ft long, 2 ft wide, and 3 ft tall. How many of these boxes can fit in a space that is 6 ft by 4 ft by 3 ft?

(A) 3   (B) 6

(C) 12   (D) 18

24. A school has 800 students. The principal surveys 50 students about their favorite subject. What is the population?

(A) The 50 surveyed students   (B) All 800 students in the school

(C) The principal   (D) The students who like math

25. Team A's scores: mean 72, MAD 3. Team B's scores: mean 74, MAD 8. Which team is more consistent?

(A) Team A because its MAD is smaller   (B) Team B because its mean is higher

(C) Team A because its mean is lower   (D) Team B because its MAD is larger

26. Two groups have the same mean but different MADs. What does this tell you?

(A) One group scored higher than the other   (B) The groups have different levels of variability

(C) The groups are identical   (D) The means were calculated incorrectly

27. The probability that an event will happen is $\frac{3}{8}$. What is the probability that the event will NOT happen? Write your answer as a fraction.

Your Answer:

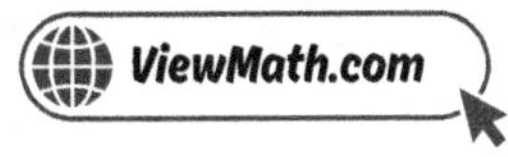

28. A bag has tiles labeled A, A, A, B, B. What is the probability of drawing a B?

(A) $\frac{1}{5}$

(B) $\frac{2}{3}$

(C) $\frac{2}{5}$

(D) $\frac{3}{5}$

29. The table shows all possible products when two dice are rolled. How many outcomes give a product of 6? Write the probability as a fraction in simplest form.

×	1	2	3	4	5	6
1	1	2	3	4	5	6
2	2	4	6	8	10	12
3	3	6	9	12	15	18
4	4	8	12	16	20	24
5	5	10	15	20	25	30
6	6	12	18	24	30	36

Your Answer:

30. Which of the following is the FIRST step in designing a simulation?

(A) Record the results

(B) Run many trials

(C) Identify the event and its possible outcomes

(D) Calculate the experimental probability

Find more at
ViewMath.com/MA-Grade7

 End of Practice Test 5 

Great job finishing the test!

 My Score

I got _____________ out of 30 questions right.

Check your answers in the **Answer Key** at the back of the book.

Review any questions you missed. That's how we learn!

Check Your Score Online!

Visit **ViewMath Academy** to enter your answers and see which topics you need to review. You can also explore lessons, take quizzes, track your scores, and save your progress!

viewmath.com/score/7.1.MA.20

Or go to viewmath.com/score and enter code: 7.1.MA.20

Practice Test 6

 30 Questions

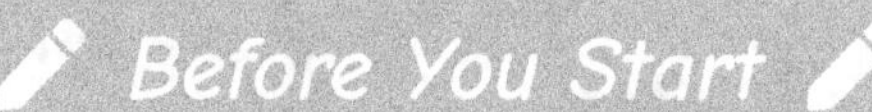

✏ Before You Start ✏

- ✓ **Read each question carefully** before choosing your answer.
- ✓ **Show your work** on scratch paper when you need to.
- ✓ **Skip hard questions** and come back to them later.
- ✓ **Check your answers** when you're done.
- ✓ **Take your time** — there's no rush!

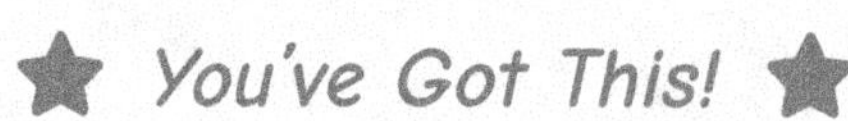

⭐ You've Got This! ⭐

Do your best and show what you know!

1. *A proportional graph passes through $(6, 15)$. What is k?*

> Your Answer:

2. *A proportional graph passes through $(4, 10)$. What is y when $x = 7$?*

(A) 13

(B) 17

(C) 17.5

(D) 28

3. *A car uses 3 gallons of gas every 87 miles. How many gallons does it need for a 435-mile trip?*

(A) 10

(B) 12

(C) 15

(D) 18

4. *45 is 75% of what number?*

(A) 33.75

(B) 50

(C) 56

(D) 60

5. *An event planner charges a 10% planning fee plus a 3% coordination fee. If an event costs $4,000, what are the total fees?*

(A) $120

(B) $400

(C) $520

(D) $1,300

Find more at
ViewMath.com/MA-Grade7

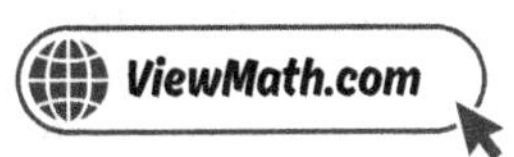

6. A scientist predicted a chemical reaction would produce 30 grams. It actually produced 28 grams. What is the percent error?

(A) 2%

(B) 6.7%

(C) 7.1%

(D) 14.3%

7. What is the opposite of the opposite of 4?

(A) -4

(B) 0

(C) 4

(D) $\frac{1}{4}$

8. Which pair of integers has a sum of -6?

(A) 4 and -10

(B) -3 and -9

(C) 8 and -2

(D) -1 and -4

9. Which expression means "a number n divided by 4, then increased by 3"?

(A) $\frac{n+3}{4}$

(B) $\frac{4}{n} + 3$

(C) $\frac{n}{4} + 3$

(D) $4n + 3$

10. Simplify $-4x + 9 + 7x - 3$.

(A) $3x + 6$

(B) $-11x + 6$

(C) $3x - 6$

(D) $11x + 12$

11. Solve $50 - 4x = 18$.

  (A) $x = -8$ (B) $x = 17$

  (C) $x = 8$ (D) $x = 12$

12. The rectangle below has an area of 56 square cm. What is the value of x?

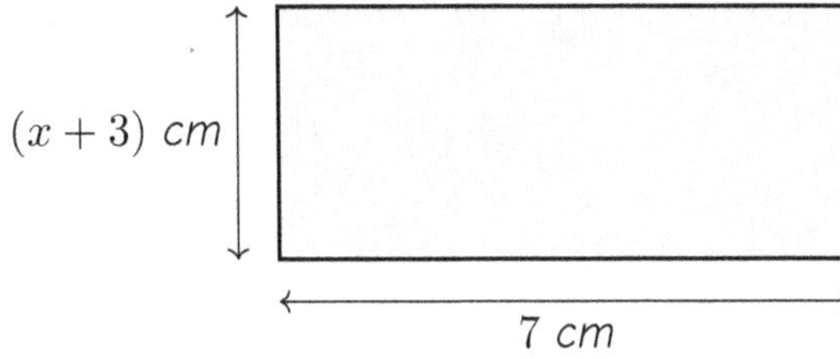

  (A) $x = 5$ (B) $x = 8$

  (C) $x = 11$ (D) $x = 4$

13. Solve $2.5x - 7.5 = 10$.

Your Answer

14. A student solved $-2x + 8 > 14$ and wrote $x > -3$. What error did the student make?

  (A) Forgot to subtract 8   (B) Divided by 2 instead of -2

  (C) Forgot to flip the inequality sign   (D) Added 8 instead of subtracting

15. Solve $5 - x > 8$ and describe the graph.

  (A) Open circle at 3, shade right   (B) Open circle at -3, shade left

  (C) Closed circle at -3, shade left   (D) Open circle at -3, shade right

16. *A drawing uses 1 cm = 6 m. It is redrawn at 1 cm = 2 m. The original drawing is 8 cm by 5 cm. What is the perimeter of the new drawing?*

  (A) *26 cm*   (B) *39 cm*

  (C) *78 cm*   (D) *120 cm*

17. *A triangle has two sides of length 8 cm and 12 cm. What is the range of possible lengths for the third side?*

> Your Answer:

18. *You are given two sides (5 cm and 7 cm) and the included angle (90°). Is the resulting triangle unique? What type of triangle is it?*

> Your Answer:

19. *You slice a cube with a vertical cut from one face to the opposite face, perpendicular to the base. What is the most likely cross-section shape?*

  (A) *Square or rectangle*   (B) *Circle*

  (C) *Triangle*   (D) *Hexagon*

20. *A circle has an area of 12.56 m². Using $\pi \approx 3.14$, what is the diameter?*

> Your Answer:

21. A rectangular room is 14 ft by 10 ft. There is a built-in closet that is 3 ft by 4 ft. What is the area of the room not including the closet?

(A) 128 ft^2

(B) 140 ft^2

(C) 152 ft^2

(D) 12 ft^2

22. A cereal box is 25 cm tall, 18 cm wide, and 5 cm deep. How much cardboard is needed to make the box?

Your Answer

23. A rectangular prism is 8 in by 5 in by 4 in. Another is 10 in by 4 in by 4 in. Which has more volume and by how much?

(A) First prism, by 10 in^3

(B) Second prism, by 10 in^3

(C) First prism, by 20 in^3

(D) They have the same volume

24. A random sample means that:

(A) Only the best members are chosen

(B) The largest possible group is chosen

(C) Every member of the population has an equal chance of being selected

(D) Only volunteers are selected

25. When is a dot plot a better choice than a box plot for comparing two groups?

(A) When data sets are very large

(B) When you want to see every individual data point

(C) When you only want to compare medians

(D) When the data is categorical

Find more at
ViewMath.com/MA-Grade7

26. Data set: $12, 15, 18, 20, 22, 25, 28$. What is the median?

(A) 18

(B) 20

(C) 22

(D) 19

27. Which of these events is impossible?

(A) Rolling a 5 on a standard die

(B) Flipping heads on a fair coin

(C) Drawing a green marble from a bag of only red and blue marbles

(D) Picking a number less than 10 from the numbers 1 to 20

28. A spinner is divided into 4 sections with probabilities: red $= 0.4$, blue $= 0.3$, green $= 0.2$, yellow $= 0.1$. What type of model is this?

(A) Uniform, because there are 4 sections

(B) Non-uniform, because the probabilities are different

(C) Invalid, because the probabilities don't add up to 1

(D) Uniform, because all sections are on the same spinner

29. A coin is flipped and a die is rolled. What is the probability of getting tails and an even number?

(A) $\frac{1}{12}$

(B) $\frac{1}{6}$

(C) $\frac{1}{4}$

(D) $\frac{1}{2}$

30. What is a simulation?

(A) A way to calculate exact probabilities

(B) A model that uses random numbers or objects to imitate a real event

(C) A method that always gives the same result

(D) A graph that shows all possible outcomes

Find more at
ViewMath.com/MA-Grade7

 # End of Practice Test 6 

Great job finishing the test!

 My Score

I got _____________ out of 30 questions right.

Check your answers in the Answer Key at the back of the book.

💡 *Review any questions you missed. That's how we learn!*

📊 Check Your Score Online!

Visit **ViewMath Academy** to enter your answers and see which topics you need to review. You can also explore lessons, take quizzes, track your scores, and save your progress!

viewmath.com/score/7.1.MA.21

Or go to viewmath.com/score and enter code: 7.1.MA.21

Practice Test 7

☑ *30 Questions*

✏ Before You Start ✏

- ✔ **Read each question carefully** before choosing your answer.
- ✔ **Show your work** on scratch paper when you need to.
- ✔ **Skip hard questions** and come back to them later.
- ✔ **Check your answers** when you're done.
- ✔ **Take your time** — there's no rush!

★ You've Got This! ★

Do your best and show what you know!

1. Find the missing value in the proportional table.

x	y
5	20
9	?

Your Answer:

2. Which statement about proportional graphs is FALSE?

(A) They always pass through the origin

(B) The slope equals the unit rate

(C) A steeper line means a larger k

(D) The line can be curved

3. A cyclist rides 42 miles in 3 hours. At this rate, how long does it take to ride 98 miles?

Your Answer:

4. 9 is 12% of what number?

(A) 1.08

(B) 36

(C) 72

(D) 75

5. An online marketplace charges a 4% seller fee. A seller lists an item for $85. How much does the seller receive after the fee?

Your Answer:

Find more at
ViewMath.com/MA-Grade7

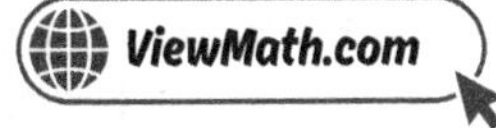

6. You estimated there were 50 marbles in a jar. The actual count was 40. What is the percent error?

(A) 10%

(B) 20%

(C) 25%

(D) 50%

7. Which integer is 5 units to the left of 2 on the number line?

Your Answer:

8. A diver is at -20 feet. She rises 8 feet. What is her new depth?

(A) -28 feet

(B) -12 feet

(C) 12 feet

(D) 28 feet

9. Which expression represents "6 more than twice a number n"?

(A) $6n + 2$

(B) $2n + 6$

(C) $2(n + 6)$

(D) $6 \times 2n$

10. Simplify $\frac{3}{4}n - \frac{1}{4}n + 5$.

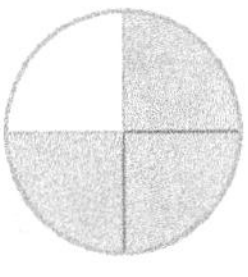

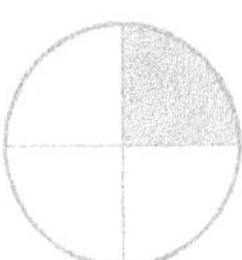

(A) $\frac{1}{2}n + 5$

(B) $n + 5$

(C) $\frac{3}{4}n + 5$

(D) $\frac{2}{4}n - 5$

11. Solve $100 - 7x = 37$.

Your Answer

12. A student solved $3(x + 4) = 30$ and wrote: $3x + 4 = 30$, $3x = 26$, $x = \frac{26}{3}$. What was the student's error?

(A) Divided by 3 instead of subtracting

(B) Did not distribute 3 to the 4

(C) Subtracted 4 from the wrong side

(D) Forgot to check the answer

13. A recipe calls for $\frac{3}{4}$ cup of sugar per batch. You already used $\frac{1}{2}$ cup. You have $2\frac{3}{4}$ cups total. The equation $\frac{3}{4}b + \frac{1}{2} = 2\frac{3}{4}$ finds how many more batches b you can make. What is b?

(A) $b = 3$

(B) $b = 4$

(C) $b = 2$

(D) $b = 3\frac{2}{3}$

Find more at
ViewMath.com/MA-Grade7

Get Online

ViewMath.com

14. *Which inequality means "no more than 25"?*

(A) $x > 25$ 　　　　　　　　(B) $x \geq 25$

(C) $x < 25$ 　　　　　　　　(D) $x \leq 25$

15. *Solve $3x + 4 > 16$. Describe the graph on a number line.*

Your Answer:

16. *A blueprint uses $1\ cm = 4\ m$. You want to redraw it at $1\ cm = 8\ m$. A wall that is $10\ cm$ on the original becomes how long on the new drawing?*

(A) 5 cm 　　　　　　　　(B) 10 cm

(C) 20 cm 　　　　　　　　(D) 40 cm

17. *Can you draw a triangle with sides 3 cm, 4 cm, and 8 cm?*

(A) Yes, and it is a unique triangle 　　　(B) Yes, and more than one triangle is possible

(C) No, because the angles don't add to 180° 　　　(D) No, because $3 + 4 < 8$

18. *Which set of measurements does NOT determine a unique triangle?*

(A) Sides 3, 4, 5 　　　　　　　　(B) Sides 7 and 9 with included angle 50°

(C) Angles 40°, 70°, 70° 　　　　　　　　(D) Angles 30°, 60° with included side 10 cm

Find more at
ViewMath.com/MA-Grade7

19. A cylinder is sliced with a vertical cut through its center, perpendicular to the base. What shape is the cross-section?

(A) Circle   (B) Rectangle

(C) Oval   (D) Triangle

20. Find the area of a semicircle with a diameter of 10 cm. Use $\pi \approx 3.14$.

Your Answer

21. A T-shaped figure can be broken into a 12 cm by 2 cm horizontal rectangle and a 2 cm by 8 cm vertical rectangle. What is the total area?

(A) 24 cm^2 (B) 32 cm^2

(C) 40 cm^2 (D) 48 cm^2

22. What does surface area measure?

(A) The space inside a 3D figure   (B) The distance around a 3D figure

(C) The total area of all faces of a 3D figure   (D) The height of a 3D figure

23. Find the volume of a rectangular prism with dimensions 9 cm by 5 cm by 4 cm.

Your Answer

 Find more at
ViewMath.com/MA-Grade7

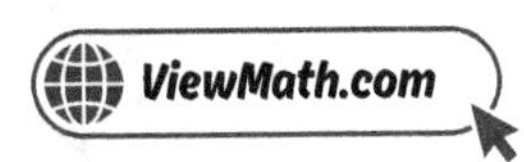

24. The diagram below shows four different sampling methods used to survey students at a school. Which method is most likely to produce an unbiased sample?

Method A
Survey the football team

Method B
Survey every 10th student on the school roster

Method C
Post survey online and wait for responses

Method D
Survey students in one math class

(A) Method A

(B) Method B

(C) Method C

(D) Method D

25. Class A: median 90, IQR 6. Class B: median 83, IQR 14. Which class is more consistent?

Your Answer:

26. Class A: median 85, IQR 8, range 20. Class B: median 80, IQR 15, range 40. A student says "Class B is better because it has a wider range." Is the student correct?

(A) Yes — wider range means more talent

(B) No — wider range means less consistency, and Class A has a higher median

(C) Yes — Class B's scores spread more, which is always better

(D) No — range does not measure anything useful

Find more at
ViewMath.com/MA-Grade7

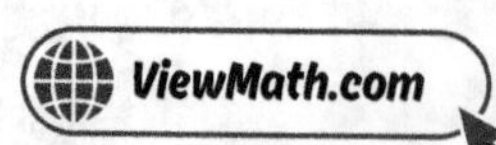

27. Look at the probability scale below. Point P is marked on the scale. A bag contains some red and blue marbles. If the probability of drawing a red marble is at point P, how many red marbles could be in a bag of 20 marbles?

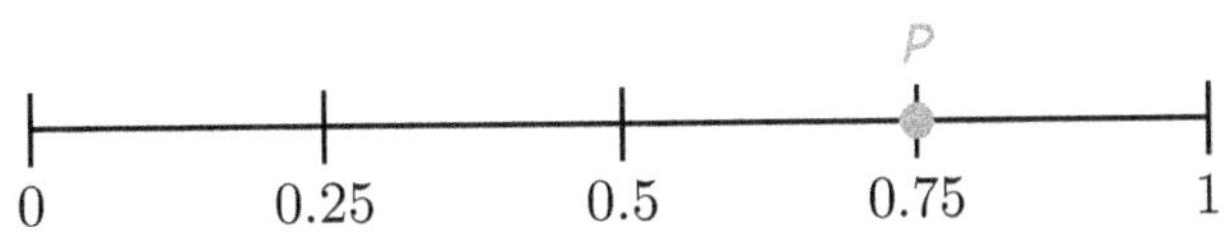

Your Answer:

28. Which of the following is an example of a uniform probability model?

(A) A bag with 3 red, 5 blue, and 2 green marbles

(B) A fair number cube with 6 sides

(C) A spinner with sections of different sizes

(D) A weighted coin

29. A coin is flipped and a die is rolled. What is the probability of getting heads and a number greater than 4?

(A) $\frac{1}{12}$

(B) $\frac{1}{6}$

(C) $\frac{2}{12}$

(D) $\frac{1}{3}$

30. A game has a 25% chance of winning. Which simulation model correctly represents this?

(A) Flip a coin — heads means win

(B) Roll a die — roll a 1 means win

(C) Use a spinner with 4 equal sections — one section means win

(D) Use digits 0–9 — digits 0–4 mean win

 # End of Practice Test 7 

Great job finishing the test!

 **My Score**

I got _____________ out of 30 questions right.

*Check your answers in the **Answer Key** at the back of the book.*

Review any questions you missed. That's how we learn!

📊 Check Your Score Online!

Visit **ViewMath Academy** to enter your answers and see which topics you need to review. You can also explore lessons, take quizzes, track your scores, and save your progress!

viewmath.com/score/7.1.MA.22

*Or go to **viewmath.com/score** and enter code: 7.1.MA.22*

Practice Test 8

☑ 30 Questions

✏ Before You Start ✏

- ✔ **Read each question carefully** before choosing your answer.
- ✔ **Show your work** on scratch paper when you need to.
- ✔ **Skip hard questions** and come back to them later.
- ✔ **Check your answers** when you're done.
- ✔ **Take your time** — there's no rush!

⭐ You've Got This! ⭐

Do your best and show what you know!

1. A car travels at a constant speed. It goes 150 miles in 2.5 hours. What is the constant of proportionality?

(A) 30 mph

(B) 60 mph

(C) 75 mph

(D) 150 mph

2. Two friends start walking at the same time. The graph shows their distances. How many miles apart are they after 6 hours?

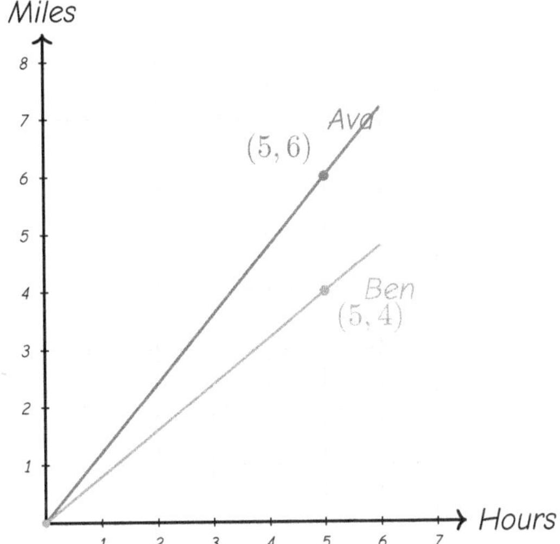

Your Answer:

3. A scale drawing uses 1 cm = 4.5 m. If a building is 27 m tall, how tall is it in the drawing?

Your Answer:

4. What is 150% of 40?

(A) 20

(B) 40

(C) 50

(D) 60

5. Estimate a 20% tip on a $93 bill. Round the bill first, then calculate.

Your Answer:

6. The number line below shows an estimated value and an actual value. What is the percent error?

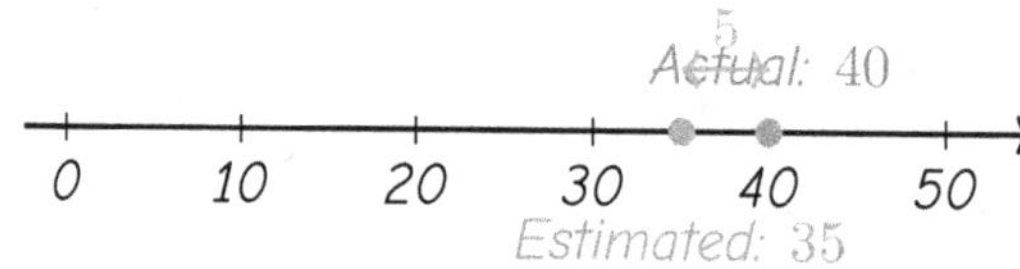

(A) 5%

(B) 10%

(C) 12.5%

(D) 14.3%

7. An elevator goes 3 floors up from the lobby, then 7 floors down. Which integer describes its final position relative to the lobby?

(A) 4

(B) −4

(C) 10

(D) −10

8. A bank account has −$50 (overdrawn). A deposit of $75 is made. What is the new balance?

(A) −$125

(B) −$25

(C) $25

(D) $125

9. Evaluate $-4y + 10$ when $y = 3$.

(A) 22

(B) −2

(C) 2

(D) −22

10. Write an expression with three terms that simplifies to $4n + 7$.

Your Answer

11. The table shows a pattern. What is the value of x when the output is 41?

Input (x)	Output
1	8
2	11
3	14
?	41

(A) $x = 12$

(B) $x = 11$

(C) $x = 13$

(D) $x = 14$

12. Solve $5(n + 3) = -10$.

(A) $n = -5$

(B) $n = 1$

(C) $n = -1$

(D) $n = -\dfrac{25}{5}$

13. Solve $\dfrac{5n}{6} + \dfrac{1}{3} = 3$.

Your Answer

14. *The table shows ticket prices for groups. A school has $200 to spend and must also pay a $35 bus fee. Write and solve an inequality to find the maximum number of students s who can go.*

Group Size	Price per Student
1–10	$12
11–25	$9
26+	$7

Assume the group qualifies for the $9 rate.

Your Answer:

15. *Solve $10 - 4x < 2$. Describe the graph.*

Your Answer:

16. *A map uses 1 cm = 15 km. A trail is 8 cm on this map. On a new map, the trail is 24 cm. What is the scale ratio from old to new?*

Your Answer:

17. *How many different rectangles have an area of 24 cm^2 if the sides must be whole numbers?*

Your Answer:

18. *Two angles of a triangle are 60° and 60°, and the side between them is 5 cm. What type of triangle is formed?*

(A) Right triangle

(B) Scalene triangle

(C) Equilateral triangle

(D) Obtuse triangle

Find more at
ViewMath.com/MA-Grade7

19. Name two different cross-section shapes you can get from slicing a cube.

Your Answer:

20. A square has a circle inscribed inside it as shown. The side of the square is 10 cm. What is the area of the circle? Use $\pi \approx 3.14$.

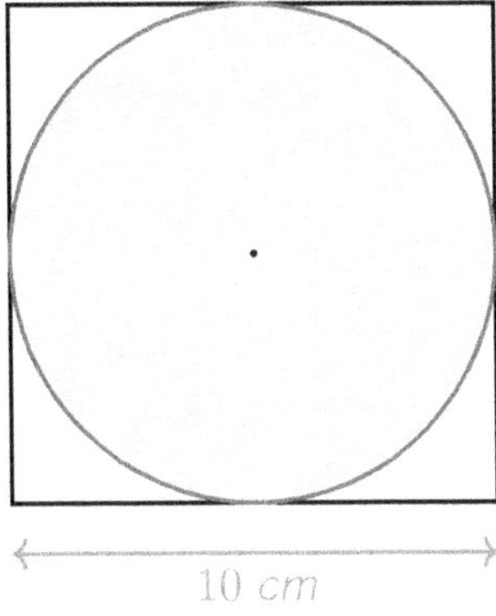

(A) 31.4 cm^2 (B) 78.5 cm^2

(C) 100 cm^2 (D) 314 cm^2

21. A rectangular patio is 12 ft by 8 ft. A circular hot tub with a diameter of 6 ft is placed on the patio. What is the area of the patio NOT covered by the hot tub? Use $\pi \approx 3.14$.

Your Answer:

22. What is the surface area of a rectangular prism with length 5 cm, width 3 cm, and height 2 cm?

(A) 30 cm^2 (B) 62 cm^2

(C) 46 cm^2 (D) 31 cm^2

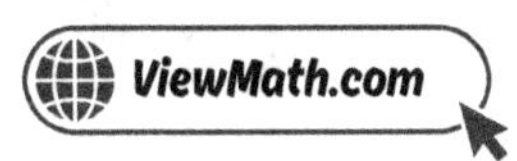

23. What is the volume of a cube with side length 5 m?

(A) $15\ m^3$

(B) $25\ m^3$

(C) $125\ m^3$

(D) $150\ m^3$

24. A random sample of 50 apples from an orchard found that 6 were bruised. If the orchard has 3,000 apples, about how many are bruised?

Your Answer:

25. Team A: mean 60, MAD 4. Team B: mean 70, MAD 3. Is the difference in means meaningful? Explain briefly.

Your Answer:

26. Data set: $20, 22, 25, 27, 30, 32, 35$. What is the IQR?

(A) 15

(B) 10

(C) 7

(D) 5

27. A spinner has 5 equal sections: red, blue, green, yellow, and white. What is the probability of landing on blue or green? Write your answer as a decimal.

Your Answer:

Find more at
ViewMath.com/MA-Grade7

28. A bag contains 5 yellow and 5 purple marbles. Lisa picks a marble, records the color, and replaces it 40 times. She gets yellow 24 times. Which statement is correct?

(A) The model predicts $P(yellow) = 0.5$, and the experiment gives $P(yellow) = 0.6$, so there is a small difference

(B) The model is wrong because $24 \neq 20$

(C) The model predicts $P(yellow) = 0.6$

(D) The experiment proves yellow is more likely than purple

29. A coin is flipped and a die is rolled. What is the probability of getting tails and a 1? Write your answer as a fraction.

Your Answer:

30. How does increasing the number of trials in a simulation affect the results?

(A) The results become less reliable

(B) The experimental probability gets closer to the theoretical probability

(C) The results stay exactly the same

(D) The simulation takes less time

Find more at
ViewMath.com/MA-Grade7

 # End of Practice Test 8 

Great job finishing the test!

☑ My Score

I got _____________ out of 30 questions right.

*Check your answers in the **Answer Key** at the back of the book.*

💡 *Review any questions you missed. That's how we learn!*

📊 Check Your Score Online!

Visit **ViewMath Academy** to enter your answers and see which topics you need to review. You can also explore lessons, take quizzes, track your scores, and save your progress!

viewmath.com/score/7.1.MA.23

Or go to viewmath.com/score and enter code: 7.1.MA.23

9

Practice Test 9

30 Questions

✏️ Before You Start ✏️

- ✔ **Read each question carefully** before choosing your answer.
- ✔ **Show your work** on scratch paper when you need to.
- ✔ **Skip hard questions** and come back to them later.
- ✔ **Check your answers** when you're done.
- ✔ **Take your time** — there's no rush!

⭐ You've Got This! ⭐

Do your best and show what you know!

1. Two students find k from the point $(8, 20)$. Student A says $k = 2.5$. Student B says $k = 0.4$. Who is correct?

(A) Student A, because $k = \frac{y}{x}$

(B) Student B, because $k = \frac{x}{y}$

(C) Both are correct, depending on which variable is independent

(D) Neither is correct

2. The equation $y = 5x$ gives the cost of movie tickets. What does the point $(1, 5)$ represent?

(A) 5 movies cost \$1

(B) 1 ticket costs \$5

(C) The total for all tickets is \$5

(D) You need 5 hours to buy 1 ticket

3. A printer prints 120 pages in 8 minutes. How long will it take to print 450 pages?

(A) 25 min

(B) 28 min

(C) 30 min

(D) 36 min

4. 14 is what percent of 56?

(A) 14%

(B) 20%

(C) 25%

(D) 28%

5. A hair stylist charges \$55 for a haircut. The customer leaves a 20% tip. What is the total the customer pays?

(A) \$11.00

(B) \$60.50

(C) \$66.00

(D) \$75.00

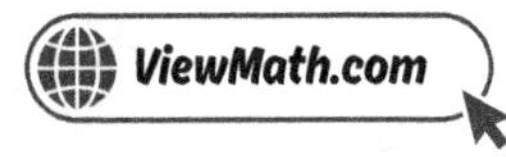

6. A builder estimated a project would cost \$8,000. The actual cost was \$7,500. What is the percent error?

(A) 5%

(B) 6.25%

(C) 6.67%

(D) 10%

7. Which expression equals 0?

(A) $7 + 7$

(B) $7 - 7$

(C) $7 \times 0 + 7$

(D) $|-7| - |7|$

8. A thermometer shows temperature changes over three hours.

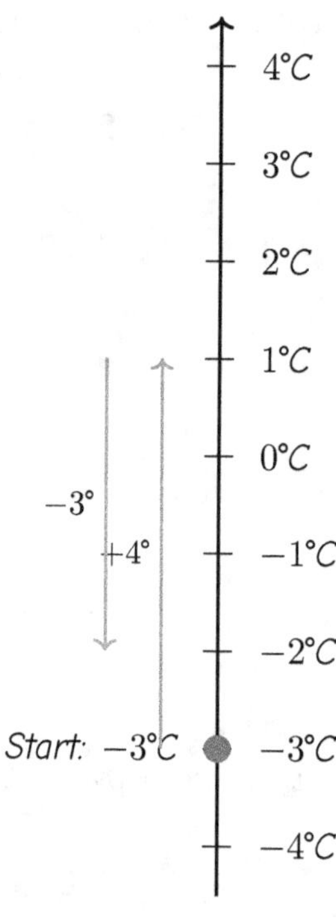

What is the final temperature after these two changes?

Your Answer:

9. *Which phrase best describes the expression* $5n - 8$?

   (A) 8 less than 5 times a number   (B) 5 less than 8 times a number

   (C) 8 minus 5 times a number   (D) 5 times a number less 8 times that number

10. *A rectangular garden has the dimensions shown. Write a simplified expression for the total distance around the garden (its perimeter).*

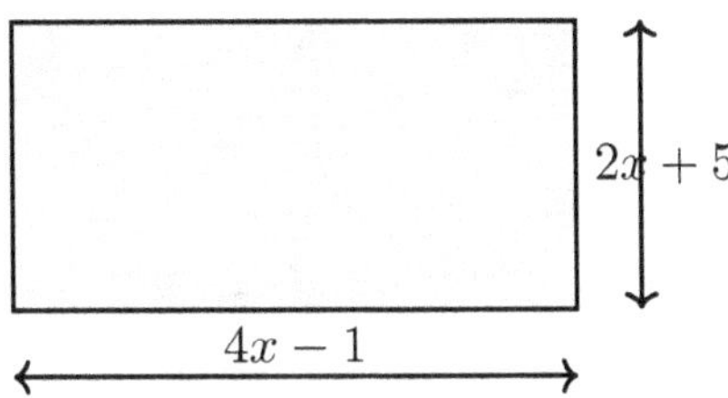

Your Answer

11. *Which equation has the solution* $x = 5$?

   (A) $4x + 3 = 18$ (B) $4x + 3 = 23$

   (C) $4x - 3 = 23$ (D) $4x - 5 = 25$

12. *Solve* $2(x + 6) = 20$.

   (A) $x = 4$ (B) $x = 7$

   (C) $x = 10$ (D) $x = 14$

13. A shirt is on sale for $\frac{2}{3}$ of its original price. After using a \$5 coupon, you pay \$15. What equation finds the original price p?

(A) $\frac{2}{3}p + 5 = 15$

(B) $p - \frac{2}{3} - 5 = 15$

(C) $\frac{2}{3}p - 5 = 15$

(D) $\frac{2}{3}(p - 5) = 15$

14. Solve $-3n + 6 \leq 0$.

(A) $n \leq 2$

(B) $n \leq -2$

(C) $n \geq 2$

(D) $n \geq -2$

15. Solve $\frac{x}{2} - 3 > 1$ and describe the graph.

(A) Open circle at 8, shade right

(B) Closed circle at 8, shade right

(C) Open circle at 4, shade right

(D) Open circle at 8, shade left

16. A blueprint uses 1 in = 12 ft. You redraw at 1 in = 4 ft. A door that is 0.5 in wide on the original becomes how wide?

(A) 0.5 in

(B) 1 in

(C) 1.5 in

(D) 3 in

17. A rectangle must have a perimeter of 24 cm. How many different rectangles can satisfy this condition?

(A) None

(B) Exactly one

(C) Exactly two

(D) More than one

Find more at
ViewMath.com/MA-Grade7

18. Two sides of a triangle are 9 cm and 12 cm with an included angle of 60°. What type of condition is this?

(A) SSS

(B) SAS

(C) ASA

(D) AAA

19. You cut a cone with a vertical slice through its apex (tip). What shape is the cross-section?

(A) Circle

(B) Triangle

(C) Rectangle

(D) Semicircle

20. A quarter-circle has a radius of 8 cm. What is its area? Use $\pi \approx 3.14$.

(A) $200.96\ cm^2$

(B) $100.48\ cm^2$

(C) $50.24\ cm^2$

(D) $25.12\ cm^2$

21. A pentagon can be divided into a rectangle 6 cm by 4 cm and a triangle with base 6 cm and height 2 cm. What is the total area?

(A) $24\ cm^2$

(B) $30\ cm^2$

(C) $36\ cm^2$

(D) $48\ cm^2$

22. What is the surface area of a cube with side length 4 cm?

(A) $16\ cm^2$

(B) $64\ cm^2$

(C) $96\ cm^2$

(D) $24\ cm^2$

Find more at
ViewMath.com/MA-Grade7

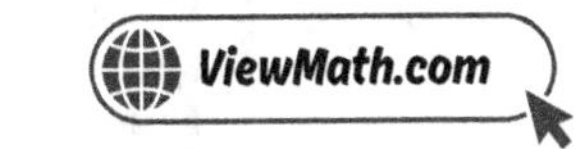

23. A rectangular prism has a base area of 24 cm^2 and a height of 7 cm. What is its volume?

(A) 31 cm^3

(B) 168 cm^3

(C) 336 cm^3

(D) 17 cm^3

24. A researcher wants to know the average height of seventh graders in a state. She measures the heights of 100 seventh graders from different schools. What is the sample?

(A) All seventh graders in the state

(B) All students in the schools she visited

(C) The 100 seventh graders she measured

(D) The tallest seventh graders

25. The box plots below compare test scores for Class A and Class B.

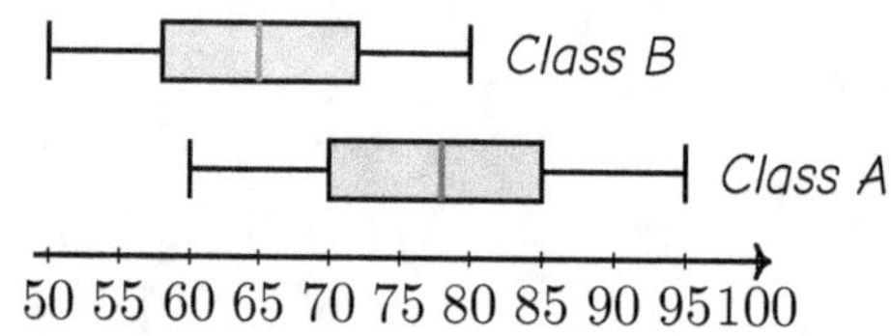

Which statement is best supported by the box plots?

(A) Class B scored higher than Class A

(B) Class A has a higher median and a larger IQR

(C) Class A has a higher median and both classes have the same IQR

(D) Class A and Class B overlap, but Class A's median is higher

26. Group A: $\{60, 65, 70, 75, 80\}$. Group B: $\{40, 55, 70, 85, 100\}$. Both have a mean of 70. Which group is more variable?

(A) Group A

(B) Group B

(C) They are equally variable

(D) Cannot be determined

Find more at
ViewMath.com/MA-Grade7

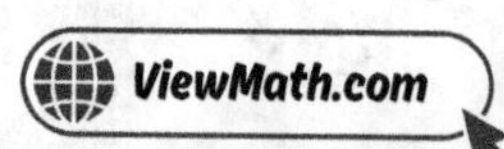

27. The bar graph below shows the contents of a bag of marbles. What is the probability of randomly drawing a blue marble?

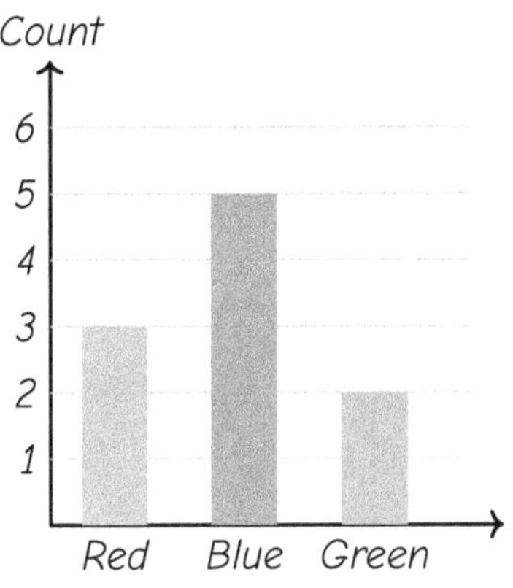

(A) $\frac{3}{10}$

(B) $\frac{1}{2}$

(C) $\frac{5}{10}$

(D) $\frac{2}{10}$

28. A teacher predicted that a spinner would land on red $\frac{1}{4}$ of the time. After 80 spins, red appeared 25 times. How do the observed and predicted results compare?

(A) They are very different — the model is wrong

(B) The observed $(\frac{25}{80} = 0.3125)$ is close to the predicted (0.25)

(C) The observed result is exactly equal to the pre-dicted result

(D) The predicted probability is higher than the observed

29. Three coins are flipped. What is the probability of getting NO heads (all tails)?

(A) $\frac{1}{2}$

(B) $\frac{1}{4}$

(C) $\frac{1}{8}$

(D) $\frac{3}{8}$

30. Describe how to use a standard die to simulate a $\frac{1}{3}$ probability event.

Your Answer:

 *End of Practice Test 9* 

Great job finishing the test!

✅ My Score

I got _____________ out of 30 questions right.

Check your answers in the **Answer Key** at the back of the book.

💡 Review any questions you missed. That's how we learn!

📊 Check Your Score Online!

Visit **ViewMath Academy** to enter your answers and see which topics you need to review. You can also explore lessons, take quizzes, track your scores, and save your progress!

viewmath.com/score/7.1.MA.24

Or go to viewmath.com/score and enter code: 7.1.MA.24

Practice Test 10

30 Questions

✏ Before You Start ✏

- ✔ **Read each question carefully** before choosing your answer.
- ✔ **Show your work** on scratch paper when you need to.
- ✔ **Skip hard questions** and come back to them later.
- ✔ **Check your answers** when you're done.
- ✔ **Take your time** — there's no rush!

⭐ You've Got This! ⭐

Do your best and show what you know!

1. The graph below shows a proportional relationship. What is the constant of proportionality?

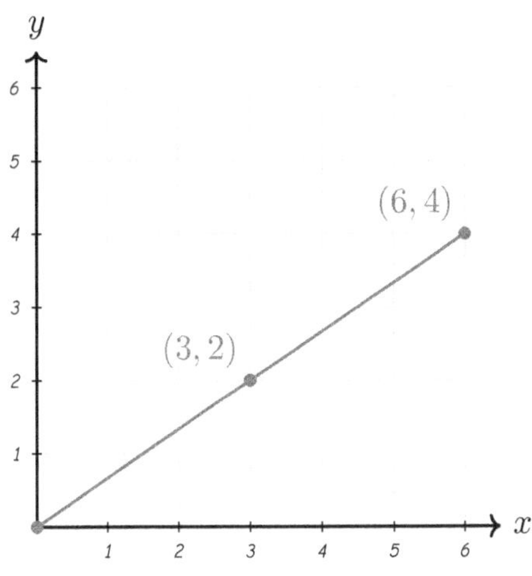

  (A) $k = \frac{1}{2}$ (B) $k = \frac{2}{3}$

  (C) $k = \frac{3}{2}$ (D) $k = 2$

2. The graph below shows a proportional relationship between hours and miles traveled. What is the unit rate?

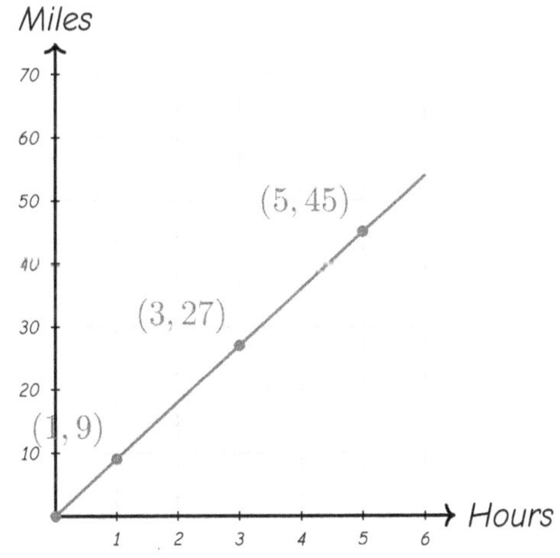

  (A) 3 mph   (B) 9 mph

  (C) 27 mph   (D) 45 mph

3. A map scale says 2 cm = 15 miles. Two cities are 7 cm apart on the map. What is the actual distance?

(A) 42.5 miles

(B) 45 miles

(C) 52.5 miles

(D) 105 miles

4. A library has 480 books. If 35% are fiction, how many fiction books does the library have?

(A) 148

(B) 160

(C) 168

(D) 172

5. Two friends split a $70 dinner bill equally and each leave a 15% tip on their share. How much does each person pay in total?

(A) $35.00

(B) $37.25

(C) $40.25

(D) $45.50

6. The bar graph shows the estimated and actual amounts of water (in liters) collected during an experiment over three days. On which day was the percent error the largest, and what was it?

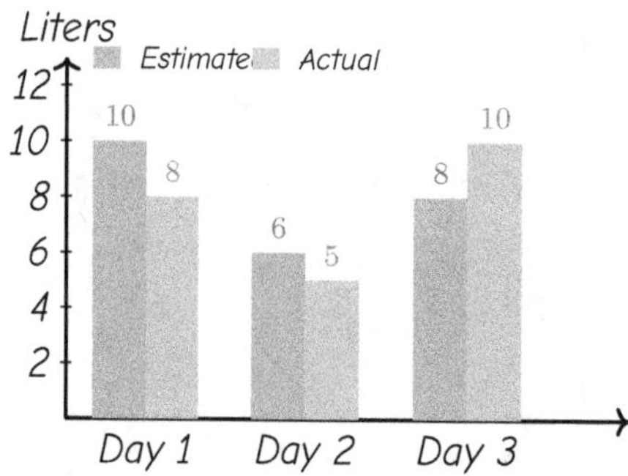

Your Answer:

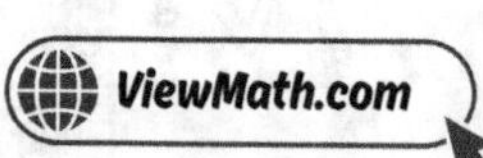

7. Find $|-23|$.

Your Answer

8. What is $(-18) + 18$?

Your Answer

9. Evaluate $2a + 7$ when $a = -3$.

(A) 13

(B) 1

(C) -13

(D) -1

10. Simplify $3y + 8 - 5y - 2$.

(A) $8y + 6$

(B) $-2y + 6$

(C) $2y - 6$

(D) $-2y + 10$

11. The bar model shows an equation. Solve for x.

45			
x	x	x	6

Your Answer

12. Solve $2(4x - 3) = 26$.

(A) $x = 3$ (B) $x = 4$

(C) $x = 2$ (D) $x = 5$

13. The bar graph shows how much three students earned. Together they earned a total of \$31.50. Student C earned $\dfrac{1}{2}$ as much as Student A. Find how much Student C earned.

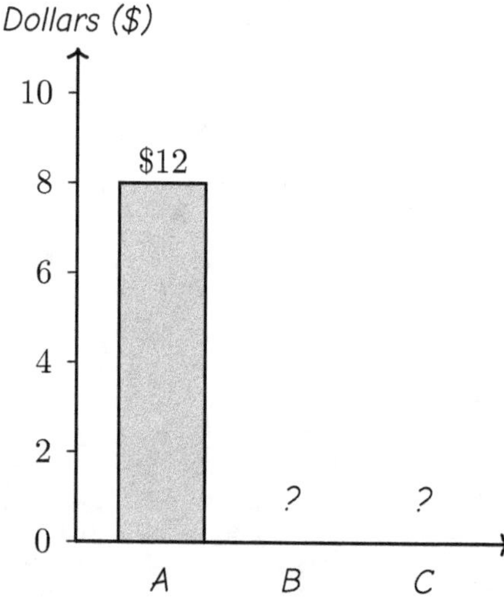

Student A earned \$12, and Student C earned $\dfrac{1}{2}$ of Student A's earnings.

Your Answer

14. Solve $-x + 9 \le 15$.

(A) $x \le -6$ (B) $x \ge 6$

(C) $x \ge -6$ (D) $x \le 6$

15. *Solve* $6 - 2x \leq -4$ *and describe the graph.*

(A) *Closed circle at 5, shade right*

(B) *Open circle at 5, shade right*

(C) *Closed circle at 5, shade left*

(D) *Closed circle at* -5*, shade right*

16. *A drawing uses* $1\ cm = 12\ m$. *You redraw at* $1\ cm = 4\ m$. *A hallway is* $5\ cm$ *on the original. How long is it on the new drawing?*

Your Answer:

17. *A circle with radius 5 cm is described as a condition. How many circles satisfy this condition?*

(A) *None*

(B) *Exactly one*

(C) *More than one (different positions)*

(D) *It depends on the center location*

18. *Look at the triangle below. Two sides and the included angle are given. What is the condition type, and is the triangle unique?*

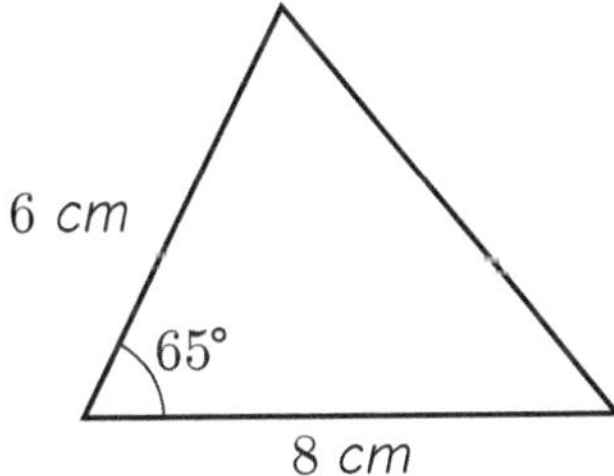

Your Answer:

19. *A cube can be sliced to produce a hexagonal cross-section. How?*

(A) *By cutting parallel to the base*

(B) *By cutting diagonally through all six faces*

(C) *By cutting vertically through the center*

(D) *It is impossible to get a hexagon from a cube*

Find more at
ViewMath.com/MA-Grade7
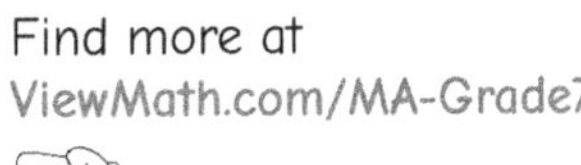

20. A pizza has a radius of 7 inches. What is the area of the pizza? Use $\pi \approx \frac{22}{7}$.

(A) 44 in^2

(B) 154 in^2

(C) 308 in^2

(D) 22 in^2

21. A composite shape is made of a rectangle that is 10 cm by 4 cm and a triangle with base 10 cm and height 3 cm attached to one side. What is the total area?

(A) 40 cm^2

(B) 55 cm^2

(C) 70 cm^2

(D) 120 cm^2

22. A cereal box is 30 cm tall, 20 cm wide, and 6 cm deep. What is its surface area?

(A) 1,440 cm^2

(B) 1,560 cm^2

(C) 1,800 cm^2

(D) 3,600 cm^2

23. A rectangular prism has a volume of 360 cm^3. Its length is 12 cm and its height is 5 cm. What is its width?

Your Answer:

Find more at
ViewMath.com/MA-Grade7

24. The table below shows the results of two samples taken from a population of 1,000 students.

	Sample 1 (50 students)	Sample 2 (50 students)
Prefer Math	18	22
Prefer Science	14	12
Prefer English	10	8
Prefer Art	8	8

Based on both samples, the best prediction for how many of the 1,000 students prefer math is:

(A) 180

(B) 220

(C) 400

(D) 360

25. Two histograms show that Group A's data is skewed left and Group B's data is symmetric. Which measure of center is most appropriate to compare them?

(A) Mean for both

(B) Median for both

(C) Mode for both

(D) Range for both

26. Data: $100, 200, 300, 400, 500$. Find the median and the IQR.

Your Answer

27. What does a probability of 1 mean?

(A) The event is impossible

(B) The event is unlikely

(C) The event has a 50% chance

(D) The event is certain to happen

28. Look at the spinner below. Which probability model correctly represents this spinner?

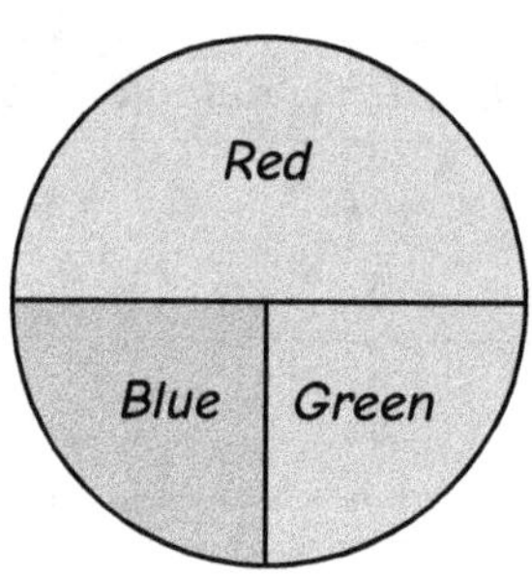

(A) $P(Red) = \frac{1}{3}$, $P(Blue) = \frac{1}{3}$, $P(Green) = \frac{1}{3}$

(B) $P(Red) = \frac{1}{2}$, $P(Blue) = \frac{1}{4}$, $P(Green) = \frac{1}{4}$

(C) $P(Red) = \frac{1}{4}$, $P(Blue) = \frac{1}{4}$, $P(Green) = \frac{1}{2}$

(D) $P(Red) = \frac{1}{2}$, $P(Blue) = \frac{1}{2}$, $P(Green) = 0$

29. Two dice are rolled. What is the probability that the product (not the sum) of the two numbers is 12?

(A) $\frac{2}{36}$

(B) $\frac{4}{36}$

(C) $\frac{3}{36}$

(D) $\frac{6}{36}$

30. A student simulates rolling two dice 36 times to see how often the sum is 7. They get a sum of 7 a total of 8 times. How does this compare to the expected number?

(A) Expected: 6; the simulation result (8) is slightly higher

(B) Expected: 6; the simulation result (8) is much higher

(C) Expected: 12; the simulation result (8) is lower

(D) Expected: 7; the simulation result (8) is exact

 *End of Practice Test 10* 

Great job finishing the test!

☑ My Score

I got _____________ out of 30 questions right.

Check your answers in the Answer Key at the back of the book.

💡 *Review any questions you missed. That's how we learn!*

📊 Check Your Score Online!

Visit **ViewMath Academy** to enter your answers and see which topics you need to review. You can also explore lessons, take quizzes, track your scores, and save your progress!

viewmath.com/score/7.1.MA.25

Or go to viewmath.com/score and enter code: 7.1.MA.25

Answer Key & Explanations

Answer Key

First try each test on your own, then check your work here.

Practice Test 1 — Answer Key

1 $k = 2.5$; missing $x = 6$ **2** B   **3** A   **4** C   **5** B   **6** $\approx 4.2\%$ **7** B   **8** C

9 6   **10** $\frac{5}{4}x - 2$ or $1\frac{1}{4}x - 2$ **11** A   **12** $t = \$14$ **13** $p = \$60$ **14** A   **15** $x \geq -6$

16 A   **17** Equilateral triangle   **18** AAA   **19** B   **20** B   **21** B   **22** A   **23** C

24 B   **25** C   **26** B   **27** A   **28** 0.4   **29** B   **30** 48

💡 Time to Learn! 💡

*Review the explanations below, **especially for the questions you missed.***

Understanding why each answer is correct builds stronger problem-solving skills.

Tip: *Circle any questions you got wrong, then read their explanation carefully.*

Practice Test 1 — Detailed Explanations

1 $k = \frac{7.5}{3} = 2.5$. For the missing row: $15 = 2.5x$ so $x = 15 \div 2.5 = 6$.

Find more at
ViewMath.com/MA-Grade7

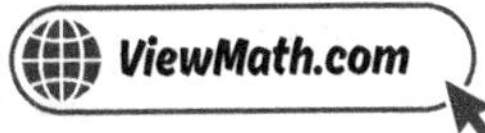

2 Line P: $k = \frac{7}{5} = 1.4$. Line Q: $k = \frac{6}{6} = 1$. Line P is steeper and has the greater unit rate.

3 Worker A: $60 \div 5 = 12$ boxes/hr. Worker B: $45 \div 5 = 9$ boxes/hr. Worker A is faster by $12 - 9 = 3$ boxes/hr.

4 $60\% = 0.60$. Multiply: $0.60 \times 300 = 180$ items.

5 Commission: $6{,}000 \times 0.05 = \$300$. Total: $400 + 300 = \$700$.

6 $\frac{|5-4.8|}{4.8} \times 100 = \frac{0.2}{4.8} \times 100 \approx 4.17 \approx 4.2\%$.

7 Point P is at -4 and Q is at 4. They are the same distance from 0 on opposite sides, so they are opposites.

8 Adding 0 to any number gives that number: $0 + (-7) = -7$.

9 Substitute: $\frac{5^2-1}{4} = \frac{25-1}{4} = \frac{24}{4} = 6$.

10 Find a common denominator: $\frac{2}{4}x + \frac{3}{4}x = \frac{5}{4}x$. The constant stays: $\frac{5}{4}x - 2$.

11 Perimeter: $(2x+1)+(x+3)+(x+5) = 39$. Combine: $4x+9 = 39$. Subtract 9: $4x = 30$. Divide by 4: $x = 7.5$.

12 $4(t+3) = 68$. Divide by 4: $t+3 = 17$. Subtract 3: $t = 14$. Each ticket costs $\$14$.

13 After $\frac{1}{4}$ markdown: $p - \frac{p}{4} = \frac{3p}{4}$. After subtracting \$8: $\frac{3p}{4} - 8 = 37$. Add 8: $\frac{3p}{4} = 45$. Multiply by $\frac{4}{3}$: $p = 60$.

14 Subtract 3: $2x > 8$. Divide by 2: $x > 4$.

15 Closed circle means -6 is included, shading right means values greater than or equal to -6: $x \geq -6$.

Find more at
ViewMath.com/MA-Grade7

16   Scale ratio: $\frac{3}{9} = \frac{1}{3}$. New length: $12 \times \frac{1}{3} = 4$ cm.

17   Two equal sides with a 60° included angle gives a triangle where all angles are 60° (isosceles with 60° forces equilateral). All three sides are 5 cm.

18   Three angles determine the shape but not the size. Infinitely many similar triangles can share the same three angle measures.

19   A plane that intersects exactly three faces of a rectangular prism creates a triangular cross-section (one edge on each of the three faces).

20 $r = 7$ cm. $A = \frac{22}{7} \times 49 = 154$ cm^2.

21   Rectangle: $12 \times 5 = 60$ cm^2. Triangle: $\frac{1}{2} \times 4 \times 5 = 10$ cm^2. Remaining: $60 - 10 = 50$ cm^2.

22   Two triangles: $2 \times \frac{1}{2} \times 6 \times 4 = 24$ cm^2. Three rectangles: $6 \times 10 + 5 \times 10 + 5 \times 10 = 60 + 50 + 50 = 160$ cm^2. Total: $24 + 160 = 184$ cm^2.

23 $V = 50 \times 30 \times 40 = 60{,}000$ cm^3.

24 $\frac{15}{60} = \frac{1}{4} = 25\%$. So 25% of $600 = 150$.

25   A box plot shows the quartiles (Q_1 and Q_3), so you can read the IQR directly as $Q_3 - Q_1$.

26   Company X has smaller IQR (1.5 vs. 4) and range (4 vs. 8), meaning delivery times are more predictable.

27   There are 5 green out of 10. $P(\text{not green}) = 1 - \frac{5}{10} = 0.5$.

28 $P(Z) = 1 - 0.15 - 0.45 = 0.40$.

Find more at
ViewMath.com/MA-Grade7

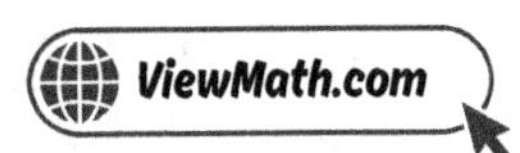

29 $P(spinner = 3) = \frac{1}{4}$, $P(die = 3) = \frac{1}{6}$. $P = \frac{1}{4} \times \frac{1}{6} = \frac{1}{24}$.

30 $80 \times 0.60 = 48$.

✅ Practice Test 2 — Answer Key

1 B

2 0.75 birdhouses per hour; 7.5 birdhouses in 10 hours

3 $37\frac{1}{2}$ cups

4 B

5 D

6 C

7 D

8 -5

9 C

10 $5a + 2b$

11 B

12 $x = -10$

13 $x = 14$

14 $x \leq 10$

15 D

16 C

17 D

18 B

19 6 cm by 4 cm

20 B

21 B

22 $156 \ cm^2$

23 C

24 4%

25 B

26 City A: mean = 66.6°F, range = 10°F. City B: mean = 68.2°F, range = 18°F. City A is more consistent.

27 C

28 B

29 C

30 B

💡 Time to Learn! 💡

Review the explanations below, **especially for the questions you missed**.

Understanding why each answer is correct builds stronger problem-solving skills.

Tip: Circle any questions you got wrong, then read their explanation carefully.

📖 Practice Test 2 — Detailed Explanations

1 $k = \frac{54}{2} = 27$, $\frac{135}{5} = 27$, $\frac{216}{8} = 27$. The car gets 27 miles per gallon.

Find more at
ViewMath.com/MA-Grade7

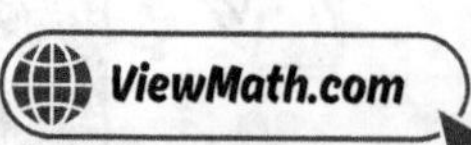

2 $k = \frac{3}{4} = 0.75$. In 10 hours: $0.75 \times 10 = 7.5$ birdhouses.

3 $k = \frac{7.5}{3} = 2.5$ cups per batch. For 15 batches: $2.5 \times 15 = 37.5 = 37\frac{1}{2}$ cups.

4 $75\% = 0.75$. Multiply: $0.75 \times 32 = 24$ students passed.

5 $17.20 \div 86 = 0.20 = 20\%$.

6   A: $\frac{|180-200|}{200} = 10\%$. B: $\frac{|220-200|}{200} = 10\%$. Both have 10% error.

7   Absolute value is the distance from 0. Since -9 is 9 units from 0, $|-9| = 9$.

8   Group positives: $5 + 3 = 8$. Group negatives: $(-11) + (-2) = -13$. Combine: $8 + (-13) = -5$.

9   Substitute: $\frac{10}{2} - 4 = 5 - 4 = 1$.

10   Combine a-terms: $-2a + 7a = 5a$. Combine b-terms: $5b - 3b = 2b$. Result: $5a + 2b$.

11   Subtract 2: $6 = \frac{x}{3}$. Multiply by 3: $x = 18$. Check: $\frac{18}{3} + 2 = 6 + 2 = 8$ ✓.

12   Divide by 2: $x + 7 = -3$. Subtract 7: $x = -10$. Check: $2(-10 + 7) = 2(-3) = -6$ ✓.

13   Subtract 1.5: $\frac{x}{4} = 3.5$. Multiply by 4: $x = 14$. The point is at 14 on the number line.

14   Subtract 7: $\frac{x}{2} \le 5$. Multiply by 2: $x \le 10$.

15   Closed circle means the endpoint is included, shading left means less than or equal to: $x \le 7$.

Find more at
ViewMath.com/MA-Grade7

16 Scale ratio: $\frac{20}{10} = 2$. Every length on the new drawing is 2 times the old one. The drawing gets bigger because each cm now represents fewer real km.

17 Three angles (AAA) determine the shape but not the size. You can draw infinitely many similar triangles of different sizes with these angles.

18 $7 + 10 = 17 > 15$ ✓, $7 + 15 = 22 > 10$ ✓, $10 + 15 = 25 > 7$ ✓. SSS with valid lengths gives one unique triangle.

19 A cut parallel to the base of a rectangular prism produces a rectangle identical to the base: 6 cm by 4 cm.

20 $r^2 = A \div \pi = 78.5 \div 3.14 = 25$. So $r = 5$ cm.

21 Rectangle: $25 \times 10 = 250$ m^2. Semicircle: $\frac{1}{2} \times 3.14 \times 5^2 = \frac{1}{2} \times 78.5 = 39.25$ m^2. Total: $250 + 39.25 = 289.25$ m^2.

22 Two triangles: $2 \times \frac{1}{2} \times 3 \times 4 = 12$ cm^2. Three rectangles: $3 \times 12 + 4 \times 12 + 5 \times 12 = 36 + 48 + 60 = 144$ cm^2. Total: $12 + 144 = 156$ cm^2.

23 $B = \frac{1}{2} \times 8 \times 3 = 12$ cm^2. $V = Bh = 12 \times 10 = 120$ cm^3.

24 $\frac{8}{200} = 0.04 = 4\%$.

25 Class 1 center ≈ 10. Class 2 center ≈ 13. Difference ≈ 3 push-ups. Class 2 is higher.

26 City A: $(62 + 64 + 64 + 66 + 66 + 66 + 68 + 68 + 70 + 72) \div 10 = 666 \div 10 = 66.6$. Range $= 72 - 62 = 10$. City B: $(58 + 60 + 64 + 66 + 68 + 70 + 72 + 74 + 74 + 76) \div 10 = 682 \div 10 = 68.2$. Range $= 76 - 58 = 18$. City A has the smaller range.

27 A probability of 0 means the event cannot happen — it is impossible.

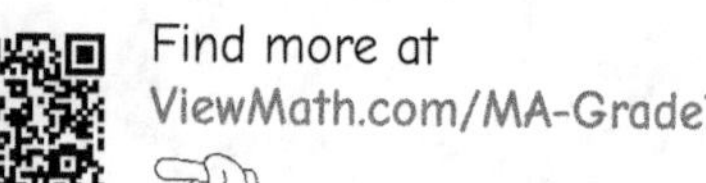

Find more at
ViewMath.com/MA-Grade7

28 Each color should appear about 50 times (200×0.25). The results $(55, 48, 52, 45)$ are all close to 50, supporting the model.

29 Outcomes with at least one tail: HT, TH, TT — that is 3 out of 4. $P = \frac{3}{4}$.

30 With 4 equal sections, each should appear about 25 times in 100 spins. The results $(26, 24, 28, 22)$ are all close to 25, suggesting the spinner is fair.

✓ Practice Test 3 — Answer Key

1 B	**2** B	**3** B	**4** C	**5** C	**6** C	**7** C	**8** D	**9** B	**10** A
11 C	**12** $x = 8$	**13** B	**14** $n \geq 5$	**15** C	**16** C	**17** B	**18** B	**19** B	
20 B	**21** B	**22** 192 cm^2	**23** C	**24** B	**25** B	**26** 2 MADs	**27** B	**28** B	
29 C	**30** A								

💡 Time to Learn! 💡

Review the explanations below, **especially for the questions you missed**.

Understanding why each answer is correct builds stronger problem-solving skills.

Tip: Circle any questions you got wrong, then read their explanation carefully.

📖 Practice Test 3 — Detailed Explanations

1 If $k = \frac{3}{4}$, then $y = \frac{3}{4}x$. For $x = 8$: $y = \frac{3}{4} \times 8 = 6$. So $(8, 6)$ works.

Find more at
ViewMath.com/MA-Grade7

2 A steeper line means a higher slope, which equals a greater unit rate (constant of proportionality).

3 Store A: $7.50 \div 3 = \$2.50$ each. Store B: $11.25 \div 5 = \$2.25$ each. Store B is cheaper.

4 Divide: $63 \div 0.90 = 70$.

5 10% of $\$78 = \$7.80$. Half of that (5%) $= \$3.90$. Add: $7.80 + 3.90 = \$11.70$.

6 $\frac{|4.5 - 5|}{5} \times 100 = \frac{0.5}{5} \times 100 = 10\%$.

7 Both $|10| = 10$ and $|-10| = 10$. Two numbers have the same absolute value: the number and its opposite.

8 A number plus its opposite (additive inverse) always equals 0.

9 Substitute: $(-2)^2 + 3(-2) = 4 + (-6) = -2$.

10 $-6y + 4y = -2y$. Then $-2y - y = -3y$.

11 Subtract 9: $3x = 21$. Divide by 3: $x = 7$. Check: $9 + 3(7) = 9 + 21 = 30$ ✓.

12 Divide by 9: $x - 3 = 5$. Add 3: $x = 8$. Check: $9(8 - 3) = 9(5) = 45$ ✓.

13 Subtract 2: $-0.4y = 4$. Divide by -0.4: $y = -10$. Check: $-0.4(-10) + 2 = 4 + 2 = 6$ ✓.

14 Subtract 10: $-5n \leq -25$. Divide by -5 and flip: $n \geq 5$.

15 Subtract 12: $-3x \geq -12$. Divide by -3 and flip: $x \leq 4$. Closed circle at 4, shade left.

Find more at
ViewMath.com/MA-Grade7

16 Scale ratio: $\frac{60}{20} = 3$. Area scales by $3^2 = 9$. New area: $4 \times 9 = 36 \ cm^2$.

17 Three side lengths that satisfy the triangle inequality (SSS) produce exactly one unique triangle. Check: $5 + 6 = 11 > 7$, $5 + 7 = 12 > 6$, $6 + 7 = 13 > 5$. All pass.

18 $90 + 45 + 45 = 180°$ ✓. But AAA alone allows different sizes. Many 45-45-90 triangles exist.

19 A vertical cut perpendicular to the triangular bases of a triangular prism produces a rectangle (the cut goes through the lateral faces).

20 $r = 8 \div 2 = 4 \ ft$. $A = \pi r^2 = 3.14 \times 16 = 50.24 \ ft^2$.

21 Triangle: $\frac{1}{2} \times 6 \times 8 = 24 \ cm^2$. Rectangle: $8 \times 3 = 24 \ cm^2$. Total: $24 + 24 = 48 \ cm^2$.

22 The prism is $10 \times 4 \times 4$. $SA = 2(10 \times 4) + 2(10 \times 4) + 2(4 \times 4) = 80 + 80 + 32 = 192 \ cm^2$.

23 $B = \frac{1}{2}(4 + 6) \times 3 = 15 \ cm^2$. $V = 15 \times 10 = 150 \ cm^3$.

24 Larger random samples better represent the population and give more reliable results.

25 Difference in medians: $88 - 82 = 6$. In MADs: $\frac{6}{4} = 1.5$ MADs.

26 Difference: $57 - 45 = 12$. In MADs: $\frac{12}{6} = 2$.

27 The letters in MATH are M, A, T, H. Only A is a vowel. $P = \frac{1}{4}$.

28 The probability of drawing red $(\frac{4}{5})$ is different from drawing blue $(\frac{1}{5})$, so this is a non-uniform model.

Find more at
ViewMath.com/MA-Grade7

29) The only way to get a sum of 2 is $(1, 1)$. $P = \frac{1}{36}$.

30) $P = \frac{6}{40} = 0.15$. The theoretical probability is $\frac{1}{8} = 0.125$, close to the simulation result.

✅ Practice Test 4 — Answer Key

1) B	2) $112.50	3) B	4) 110
5) C	6) C	7) C	8) 15 feet
9) C	10) A	11) A	12) $x = 5$
13) A	14) $x \leq -5$	15) C	16) C
17) B	18) C	19) C	20) D
21) B	22) C	23) B	24) Random variation (or the sample may be biased)
25) C	26) B	27) C	28) C
29) $\frac{3}{8}$	30) B		

💡 Time to Learn! 💡

Review the explanations below, **especially for the questions you missed**.

Understanding why each answer is correct builds stronger problem-solving skills.

Tip: Circle any questions you got wrong, then read their explanation carefully.

📖 Practice Test 4 — Detailed Explanations

1) $k = \frac{y}{x} = \frac{12}{3} = 4$. Check: $\frac{20}{5} = 4$ and $\frac{32}{8} = 4$.

2) $k = \frac{67.50}{3} = 22.50$ per hour. Cost for 5 hours: $22.50 \times 5 = \$112.50$.

Find more at
ViewMath.com/MA-Grade7

3 Factory 1: $168 \div 7 = 24$ items/hr. Factory 2: $200 \div 8 = 25$ items/hr. Factory 2 is more productive.

4 $44\% = 0.44$. Multiply: $0.44 \times 250 = 110$ chocolate candies.

5 $15\% = 0.15$. Tip: $40 \times 0.15 = \$6.00$.

6 Absolute value ensures the percent error is positive whether the estimate was too high or too low.

7 Absolute value measures distance from 0, which is always ≥ 0. It can never be negative.

8 $(-30) + 45 = 15$. She is now 15 feet above sea level.

9 "The product of 7 and p" is $7p$. "Decreased by 12" means subtract 12: $7p - 12$.

10 $7x$ and $3x$ are like terms. Add the coefficients: $7 + 3 = 10$. Result: $10x$.

11 Subtract 12: $\frac{w}{5} = 8$. Multiply by 5: $w = 40$. Check: $\frac{40}{5} + 12 = 8 + 12 = 20$ ✓.

12 Divide by 5: $2x - 3 = 7$. Add 3: $2x = 10$. Divide by 2: $x = 5$. Check: $5(2(5) - 3) = 5(7) = 35$ ✓.

13 If $\frac{2}{3}x = 8$, then each $\frac{1}{3}$ of x is 4. So $x = 3 \times 4 = 12$.

14 Subtract 8: $-3x \geq 15$. Divide by -3 and flip: $x \leq -5$.

15 Divide by -2 and flip: $x \geq -5$. Closed circle at -5, shade right.

16 Scale ratio: $\frac{4}{2} = 2$. New base: $5 \times 2 = 10$ cm.

Find more at
ViewMath.com/MA-Grade7

17 AAA with a specified side length fixes the size. A 60°-60°-60° triangle is equilateral, so all sides are 4 cm. Exactly one triangle.

18 Although $90 + 100 + (-10) = 180$, every angle in a triangle must be greater than 0°. A negative angle is impossible.

19 A rectangular prism has flat faces and straight edges. Its cross-sections can be triangles, rectangles, pentagons, or hexagons, but never circles (since it has no curved surfaces).

20 $A = \pi r^2 = 3.14 \times 100 = 314 \ cm^2$.

21 Square: $8^2 = 64 \ cm^2$. Semicircle: $\frac{1}{2} \times \pi r^2 = \frac{1}{2} \times 3.14 \times 16 = 25.12 \ cm^2$. Total: $64 + 25.12 = 89.12 \ cm^2$.

22 The three pairs of faces have areas $7 \times 4 = 28$, $7 \times 3 = 21$, and $4 \times 3 = 12 \ m^2$. The largest is $28 \ m^2$.

23 Bottom row: 5 cubes. Second row: 3 cubes. Third row: 3 cubes. Total: $5 + 3 + 3 = 11 \ cm^3$.

24 Even a well-chosen random sample may differ slightly from the population due to chance.

25 A complete comparison requires looking at both center (where the data clusters) and spread (how spread out it is).

26 Difference $= \$620 - \$500 = \$120$. In MADs: $\frac{120}{40} = 3$ MADs.

27 $\frac{3}{4} = 0.75$, which is close to 1. The team is likely to win, but since the probability is not 1, it is not certain.

28 $P(tails) = 1 - 0.60 = 0.40$. Since heads and tails have different probabilities, this is a non-uniform model.

29 Outcomes with exactly 2 tails: HTT, THT, TTH — that is 3 out of 8. $P = \frac{3}{8}$.

Find more at
ViewMath.com/MA-Grade7

30 Each coin flip has a 50% chance of heads (pass). Flipping 3 coins and counting exactly 2 heads simulates exactly 2 passes.

✅ Practice Test 5 — Answer Key

 **1** $k = 70$ widgets per hour    **2** B    **3** B    **4** D    **5** C    **6** $\approx 7.1\%$  **7** C

 **8** C    **9** B   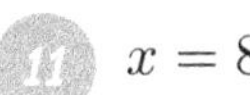 **10** $10x + 5$  **11** $x = 8$  **12** B    **13** $x = 8$  **14** A    **15** B

 **16** $96\ cm^2$ 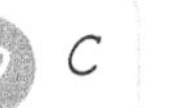 **17** D    **18** C    **19** C    **20** C    **21** $52\ cm^2$  **22** C    **23** B    **24** B

 **25** A    **26** B    **27** $\frac{5}{8}$  **28** C   **29** $\frac{1}{9}$ **30** C

💡 Time to Learn! 💡

Review the explanations below, **especially for the questions you missed**.

Understanding why each answer is correct builds stronger problem-solving skills.

Tip: Circle any questions you got wrong, then read their explanation carefully.

📖 Practice Test 5 — Detailed Explanations

1 $k = \frac{175}{2.5} = 70$ widgets per hour.

2 Unit rate $= k = \frac{18}{3} = 6$.

3 12-oz: $3.60 \div 12 = \$0.30$/oz. 20-oz: $5.40 \div 20 = \$0.27$/oz. The 20-oz box is the better value.

Find more at
ViewMath.com/MA-Grade7

4 The shaded portion is $3 \times 15 = 45$. The total is 75. Percent: $45 \div 75 = 0.60 = 60\%$.

5 $250{,}000 \times 0.03 = \$7{,}500$.

6 $\frac{|45-42|}{42} \times 100 = \frac{3}{42} \times 100 \approx 7.14 \approx 7.1\%$.

7 On a number line, -7 is to the left of -4, so $-7 < -4$.

8 $(-9) + n = -2$ means $n = -2 - (-9) = -2 + 9 = 7$.

9 "The sum of k and 10" is $k + 10$. "Half" of that sum is $\frac{k+10}{2}$.

10 Combine variable terms: $3x + 5x + 2x = 10x$. Combine constants: $2 - 1 + 4 = 5$. Perimeter: $10x + 5$.

11 Subtract 15: $2x = 16$. Divide by 2: $x = 8$. Check: $2(8) + 15 = 31$ ✓.

12 Divide by -4: $y + 3 = -7$. Subtract 3: $y = -10$. Check: $-4(-10 + 3) = -4(-7) = 28$ ✓.

13 Subtract 4.2: $-0.3x = -2.4$. Divide by -0.3: $x = 8$. Check: $-0.3(8) + 4.2 = -2.4 + 4.2 = 1.8$ ✓.

14 Current height plus growth: $42 + g > 48$.

15 Open circle means -1 is not included, shading left means less than: $x < -1$.

16 Scale ratio: $\frac{80}{20} = 4$. New dimensions: $2 \times 4 = 8$ cm and $3 \times 4 = 12$ cm. New area: $8 \times 12 = 96$ cm^2.

17 Two side lengths of a parallelogram do not fix the angles. The shape can be a rectangle, a rhomboid, or anything in between. Many different parallelograms are possible.

Find more at
ViewMath.com/MA-Grade7

18 The three angles sum to 190°, not 180°. Triangle angles must always sum to exactly 180°.

19 A cone has a circular base, and all horizontal cuts parallel to the base produce circles of varying sizes.

20 $A = \pi r^2 = 3.14 \times 25 = 78.5\ m^2$.

21 Top: $10 \times 2 = 20\ cm^2$. Bottom: $10 \times 2 = 20\ cm^2$. Middle: $2 \times 6 = 12\ cm^2$. Total: $20 + 20 + 12 = 52\ cm^2$.

22 A rectangular prism has 6 faces: top, bottom, front, back, left, and right.

23 Space volume: $6 \times 4 \times 3 = 72\ ft^3$. Box volume: $2 \times 2 \times 3 = 12\ ft^3$. Number: $72 \div 12 = 6$ boxes.

24 The population is the entire group you want to study — all 800 students.

25 A smaller MAD means scores are closer to the mean, so Team A is more consistent.

26 Same mean means same center. Different MADs mean one group's data is more spread out than the other's.

27 $P(not\ happening) = 1 - \frac{3}{8} = \frac{5}{8}$.

28 There are 2 B tiles out of 5 total. $P(B) = \frac{2}{5}$.

29 Products of 6: $(1,6), (2,3), (3,2), (6,1)$ — 4 outcomes. $P = \frac{4}{36} = \frac{1}{9}$.

30 Before choosing a model or running trials, you must first identify what event you're simulating and what the possible outcomes are.

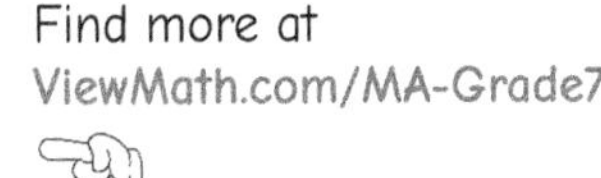
Find more at
ViewMath.com/MA-Grade7

📋 Practice Test 6 — Answer Key

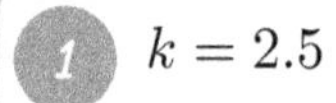 $k = 2.5$  C    C    D    C    C    C    A    C

 A   **11** C   **12** A    $x = 7$ **14** C   **15** B   **16** C

17 Greater than 4 cm and less than 20 cm   **18** Yes, unique; it is a right triangle   **19** A   **20** 4 m

21 A   **22** 1330 cm^2 **23** D   **24** C   **25** B   **26** B   **27** C   **28** B   **29** C

30 B

💡 Time to Learn! 💡

Review the explanations below, **especially for the questions you missed**.

Understanding why each answer is correct builds stronger problem-solving skills.

Tip: Circle any questions you got wrong, then read their explanation carefully.

📖 Practice Test 6 — Detailed Explanations

1 $k = \frac{y}{x} = \frac{15}{6} = 2.5$.

2 $k = \frac{10}{4} = 2.5$. When $x = 7$: $y = 2.5 \times 7 = 17.5$.

3 $\frac{3}{87} = \frac{x}{435}$. Cross-multiply: $87x = 1{,}305$, so $x = 15$ gallons.

4 Divide the part by the percent: $45 \div 0.75 = 60$.

Find more at
ViewMath.com/MA-Grade7

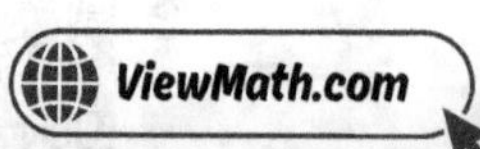

5 Planning fee: $4{,}000 \times 0.10 = \$400$. Coordination fee: $4{,}000 \times 0.03 = \$120$. Total: $400 + 120 = \$520$.

6 $\frac{|30-28|}{28} \times 100 = \frac{2}{28} \times 100 \approx 7.1\%$.

7 The opposite of 4 is -4. The opposite of -4 is 4. Taking the opposite twice returns you to the original number.

8 $4 + (-10) = -6$. Check the others: $(-3) + (-9) = -12$; $8 + (-2) = 6$; $(-1) + (-4) = -5$.

9 "Divided by 4" gives $\frac{n}{4}$. "Increased by 3" means add 3: $\frac{n}{4} + 3$.

10 Combine variable terms: $-4x + 7x = 3x$. Combine constants: $9 - 3 = 6$. Result: $3x + 6$.

11 Subtract 50: $-4x = -32$. Divide by -4: $x = 8$. Check: $50 - 4(8) = 50 - 32 = 18$ ✓.

12 Area $= 7(x + 3) = 56$. Divide by 7: $x + 3 = 8$. Subtract 3: $x = 5$.

13 Add 7.5: $2.5x = 17.5$. Divide by 2.5: $x = 7$. Check: $2.5(7) - 7.5 = 17.5 - 7.5 = 10$ ✓.

14 Correct steps: subtract 8 ⊠ $-2x > 6$. Divide by -2 and flip: $x < -3$. The student forgot to flip.

15 Subtract 5: $-x > 3$. Multiply by -1 and flip: $x < -3$. Open circle at -3, shade left.

16 Scale ratio: $\frac{6}{2} = 3$. New dimensions: $8 \times 3 = 24$ cm and $5 \times 3 = 15$ cm. New perimeter: $2(24 + 15) = 78$ cm.

17 By the triangle inequality: $12 - 8 <$ third side $< 12 + 8$, so $4 <$ third side < 20.

18 SAS produces a unique triangle. The included angle of $90°$ makes it a right triangle with legs 5 cm and 7 cm.

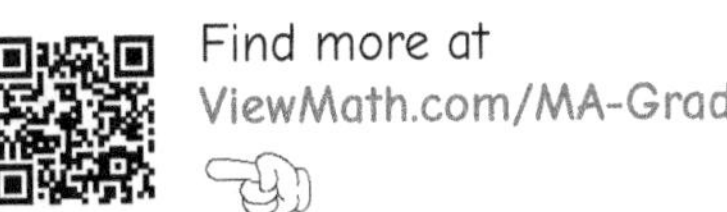
Find more at
ViewMath.com/MA-Grade7

19 A vertical cut through a cube perpendicular to the base typically produces a rectangle. If the cut goes through the center parallel to a face, it can be a square.

20 $r^2 = 12.56 \div 3.14 = 4$. So $r = 2$ m and $d = 4$ m.

21 Room: $14 \times 10 = 140$ ft^2. Closet: $3 \times 4 = 12$ ft^2. Remaining: $140 - 12 = 128$ ft^2.

22 $SA = 2(25 \times 18) + 2(25 \times 5) + 2(18 \times 5) = 900 + 250 + 180 = 1330$ cm^2.

23 First: $8 \times 5 \times 4 = 160$ in^3. Second: $10 \times 4 \times 4 = 160$ in^3. They have the same volume.

24 In a random sample, every member of the population has an equal chance of being chosen.

25 Dot plots show every data point, which is useful for small data sets where individual values matter.

26 There are 7 values. The middle value (4th) is 20.

27 If the bag contains only red and blue marbles, drawing a green marble is impossible ($P = 0$).

28 The probabilities are $0.4, 0.3, 0.2, 0.1$ — all different, so this is non-uniform. They sum to 1.0, making it valid.

29 $P(\text{tails}) = \frac{1}{2}$ and $P(\text{even}) = \frac{3}{6} = \frac{1}{2}$. $P = \frac{1}{2} \times \frac{1}{2} = \frac{1}{4}$. Or: 3 favorable out of 12 total outcomes.

30 A simulation uses random numbers, coins, dice, or spinners to model a real-world event and estimate probabilities.

☑ Practice Test 7 — Answer Key

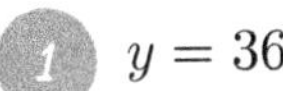 $y = 36$  D   3 7 hours    D   5 $81.60   6 C    -3  B

9 B   10 A   11 $x = 9$ 12 B   13 A   14 D   15 $x > 4$; open circle at 4, shade right

16 A   17 D   18 C   19 B   20 $39.25\ cm^2$ 21 C   22 C   23 $180\ cm^3$ 24 B

25 Class A   26 B   27 15   28 B   29 B   30 C

💡 Time to Learn! 💡

Review the explanations below, **especially for the questions you missed**.

Understanding why each answer is correct builds stronger problem-solving skills.

Tip: Circle any questions you got wrong, then read their explanation carefully.

📖 Practice Test 7 — Detailed Explanations

1 $k = \frac{20}{5} = 4$. So $y = 4 \times 9 = 36$.

2 A proportional relationship always graphs as a straight line through the origin. It cannot be curved.

3 Rate: $42 \div 3 = 14$ mph. Time: $98 \div 14 = 7$ hours.

4 Divide the part by the percent: $9 \div 0.12 = 75$.

5 Fee: $85 \times 0.04 = \$3.40$. Seller receives: $85 - 3.40 = \$81.60$.

6 $\frac{|50-40|}{40} \times 100 = \frac{10}{40} \times 100 = 25\%$.

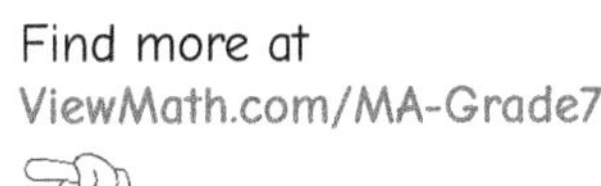

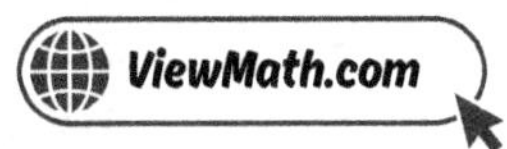

7. Moving 5 units left from 2: $2 - 5 = -3$.

8. $(-20) + 8 = -12$. Rising from a negative position adds a positive: $20 - 8 = 12$, keep negative.

9. "Twice a number" is $2n$. "6 more than" means add 6: $2n + 6$.

10. Combine: $\frac{3}{4}n - \frac{1}{4}n = \frac{2}{4}n = \frac{1}{2}n$. Add the constant: $\frac{1}{2}n + 5$.

11. Subtract 100: $-7x = -63$. Divide by -7: $x = 9$. Check: $100 - 7(9) = 100 - 63 = 37$ ✓.

12. The student wrote $3x + 4$ instead of $3x + 12$. Distributing correctly gives $3x + 12 = 30$.

13. Subtract $\frac{1}{2}$: $\frac{3}{4}b = 2\frac{1}{4} = \frac{9}{4}$. Multiply by $\frac{4}{3}$: $b = \frac{9}{4} \times \frac{4}{3} = 3$.

14. "No more than 25" means 25 or fewer: $x \leq 25$.

15. Subtract 4: $3x > 12$. Divide by 3: $x > 4$. Open circle at 4, shade right.

16. Scale ratio: $\frac{4}{8} = \frac{1}{2}$. New length: $10 \times \frac{1}{2} = 5$ cm. A larger scale (more real units per cm) shrinks the drawing.

17. The triangle inequality requires the sum of any two sides to be greater than the third. Here $3 + 4 = 7 < 8$, so no triangle can be formed.

18. AAA alone does not fix the size of the triangle. Many triangles can share these angle measures.

19. A vertical cut through the center of a cylinder produces a rectangle. The width equals the diameter and the height equals the cylinder's height.

Find more at
ViewMath.com/MA-Grade7

20. $r = 5$ cm. Full circle: $3.14 \times 25 = 78.5$ cm^2. Semicircle: $78.5 \div 2 = 39.25$ cm^2.

21. Horizontal: $12 \times 2 = 24$ cm^2. Vertical: $2 \times 8 = 16$ cm^2. Total: $24 + 16 = 40$ cm^2.

22. Surface area is the total area covering the outside of a 3D object — the sum of the areas of all its faces.

23. $V = 9 \times 5 \times 4 = 180$ cm^3.

24. Selecting every 10th student from the school roster is systematic and gives all students a chance, making it the least biased.

25. Class A has a smaller IQR (6 vs. 14), meaning the middle 50% of scores are more tightly clustered.

26. A wider range indicates more variability, not better performance. Class A has a higher median and is more consistent (smaller IQR and range).

27. Point P is at 0.75. If $P(\text{red}) = 0.75$, then $0.75 \times 20 = 15$ red marbles.

28. A fair number cube has 6 equally likely outcomes, making it a uniform model.

29. $P(\text{heads}) = \frac{1}{2}$. Numbers greater than 4: 5, 6, so $P = \frac{2}{6} = \frac{1}{3}$. $P = \frac{1}{2} \times \frac{1}{3} = \frac{1}{6}$.

30. A spinner with 4 equal sections has $\frac{1}{4} = 25\%$ chance for each section. Assigning one section as "win" gives 25%.

☑ Practice Test 8 — Answer Key

 B  2.4 miles  6 cm  D  About $18.60  C  B  C

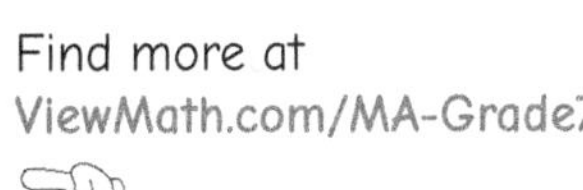

9 B

10 Answers vary, e.g., $n + 3n + 7$ or $5n - n + 7$

11 A

12 A

13 $n = \dfrac{16}{5}$ or $3\dfrac{1}{5}$

14 $9s + 35 \leq 200$; $s \leq 18$

15 $x > 2$; open circle at 2, shade right

16 3

17 4

18 C

19 Any two of: square, rectangle, triangle, pentagon, hexagon

20 B

21 67.74 ft^2

22 B

23 C

24 360

25 Yes; the means are about 2.5 to 3.3 MADs apart

26 B

27 0.4

28 A

29 $\dfrac{1}{12}$

30 B

💡 Time to Learn! 💡

Review the explanations below, **especially for the questions you missed**.

Understanding why each answer is correct builds stronger problem-solving skills.

Tip: Circle any questions you got wrong, then read their explanation carefully.

📖 Practice Test 8 — Detailed Explanations

1 $k = \dfrac{150}{2.5} = 60$ miles per hour.

2 Ava: $k = \dfrac{6}{5} = 1.2$ mph; at 6 hr: $1.2 \times 6 = 7.2$ mi. Ben: $k = \dfrac{4}{5} = 0.8$ mph; at 6 hr: $0.8 \times 6 = 4.8$ mi. Difference: $7.2 - 4.8 = 2.4$ miles.

3 $\dfrac{1}{4.5} = \dfrac{x}{27}$. Cross-multiply: $4.5x = 27$, so $x = 6$ cm.

4 $150\% = 1.50$. Multiply: $1.50 \times 40 = 60$. A percent greater than 100% gives a result greater than the whole.

5 Round to \$93. Find 10%: \$9.30. Double for 20%: $9.30 \times 2 = \$18.60$.

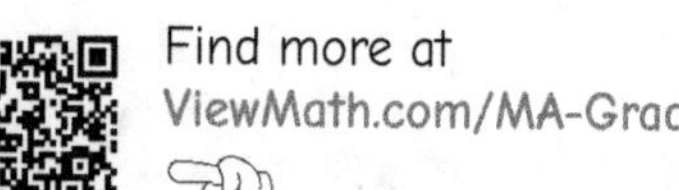

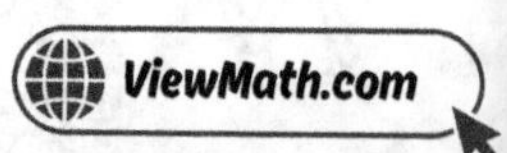

6 $\frac{|35-40|}{40} \times 100 = \frac{5}{40} \times 100 = 12.5\%$.

7 Up 3 floors, then down 7: $3 + (-7) = -4$. The elevator is 4 floors below the lobby.

8 $(-50) + 75 = 25$. The deposit is larger than the overdraft, so the balance becomes positive.

9 Substitute: $-4(3) + 10 = -12 + 10 = -2$.

10 Any expression whose like terms combine to give $4n + 7$ is correct. For example: $n + 3n + 7 = 4n + 7$.

11 The pattern is $3x + 5$. Solve $3x + 5 = 41$: subtract 5 to get $3x = 36$, divide by 3: $x = 12$. Check: $3(12) + 5 = 41$ ✓.

12 Divide by 5: $n + 3 = -2$. Subtract 3: $n = -5$. Check: $5(-5 + 3) = 5(-2) = -10$ ✓.

13 Multiply every term by 6: $5n + 2 = 18$. Subtract 2: $5n = 16$. Divide by 5: $n = \frac{16}{5} = 3\frac{1}{5}$.

14 Total cost: $9s + 35 \leq 200$. Subtract 35: $9s \leq 165$. Divide by 9: $s \leq 18.\overline{3}$. Since s must be a whole number, $s \leq 18$.

15 Subtract 10: $-4x < -8$. Divide by -4 and flip: $x > 2$. Open circle at 2, shade right.

16 $24 \div 8 = 3$. Every old measurement is multiplied by 3 on the new map.

17 Factor pairs of 24: 1×24, 2×12, 3×8, 4×6. That is 4 different rectangles.

18 The third angle is $180° - 60° - 60° = 60°$. All angles are $60°$, so the triangle is equilateral with all sides 5 cm.

Find more at
ViewMath.com/MA-Grade7

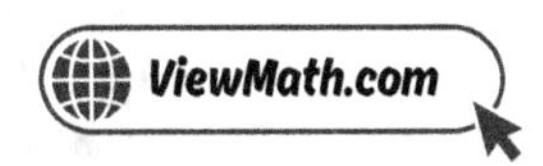

19 A cube can be sliced to produce squares, rectangles, triangles, pentagons, or hexagons depending on the angle and position of the cut.

20 The circle's diameter equals the side of the square: $d = 10$, so $r = 5$ cm. $A = \pi r^2 = 3.14 \times 25 = 78.5$ cm^2.

21 Patio: $12 \times 8 = 96$ ft^2. Hot tub: $\pi r^2 = 3.14 \times 9 = 28.26$ ft^2. Remaining: $96 - 28.26 = 67.74$ ft^2.

22 $SA = 2(5 \times 3) + 2(5 \times 2) + 2(3 \times 2) = 30 + 20 + 12 = 62$ cm^2.

23 $V = s^3 = 5^3 = 125$ m^3.

24 $\frac{6}{50} = 12\%$. Then 12% of $3{,}000 = 360$.

25 Difference $= 10$. Using MAD of 4: $\frac{10}{4} = 2.5$. Using MAD of 3: $\frac{10}{3} \approx 3.3$. Both are greater than 2, so the difference is meaningful.

26 $Q_1 = 22$, $Q_3 = 32$. $IQR = 32 - 22 = 10$.

27 2 favorable sections (blue, green) out of 5 total. $P = \frac{2}{5} = 0.4$.

28 Predicted: $\frac{5}{10} = 0.5$. Experimental: $\frac{24}{40} = 0.6$. The small difference is expected from natural variation.

29 $P(\text{tails}) = \frac{1}{2}$, $P(1) = \frac{1}{6}$. $P = \frac{1}{2} \times \frac{1}{6} = \frac{1}{12}$.

30 More trials produce more reliable estimates. The experimental probability approaches the theoretical value as trials increase.

Find more at
ViewMath.com/MA-Grade7

✅ Practice Test 9 — Answer Key

| 1 A | 2 B | 3 C | 4 C | 5 C | 6 C | 7 B | 8 −2°C | 9 A |

| 10 $12x + 8$ | 11 B | 12 A | 13 C | 14 C | 15 A | 16 C | 17 D | 18 B |

| 19 B | 20 C | 21 B | 22 C | 23 B | 24 C | 25 D | 26 B | 27 C | 28 B |

29 C   30 Roll the die; let 1 and 2 represent the event (or any 2 numbers out of 6)

💡 Time to Learn! 💡

Review the explanations below, **especially for the questions you missed**.

Understanding why each answer is correct builds stronger problem-solving skills.

Tip: Circle any questions you got wrong, then read their explanation carefully.

📖 Practice Test 9 — Detailed Explanations

1. By convention, $k = \frac{y}{x} = \frac{20}{8} = 2.5$. Student B computed $\frac{x}{y}$, which is the reciprocal — not k.

2. The point $(1, 5)$ means $x = 1$ ticket and $y = \$5$, so one ticket costs $5. This is the unit rate.

3. Rate: $120 \div 8 = 15$ pages/min. Time for 450: $450 \div 15 = 30$ min.

4. Divide the part by the whole: $14 \div 56 = 0.25 = 25\%$.

5. $55 \times 1.20 = \$66.00$.

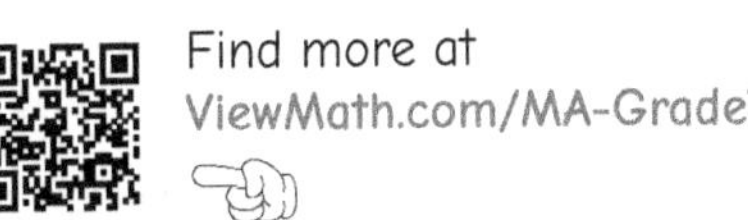
Find more at
ViewMath.com/MA-Grade7

6. $\frac{|8,000-7,500|}{7,500} \times 100 = \frac{500}{7,500} \times 100 \approx 6.67\%$.

7. $7 - 7 = 0$. Note that $|-7| - |7| = 7 - 7 = 0$ also works, but since only one answer is expected, B is the most direct additive-inverse example. Actually D also equals 0. But B is the clearest demonstration of opposites.

8. Start at $-3°C$. Rise 4°: $-3 + 4 = 1°C$. Drop 3°: $1 + (-3) = -2°C$.

9. $5n$ means "5 times a number." Subtracting 8 means "8 less than" that product.

10. Perimeter $= 2(4x - 1) + 2(2x + 5) = 8x - 2 + 4x + 10 = 12x + 8$.

11. Substitute $x = 5$: $4(5) + 3 = 20 + 3 = 23$. So $4x + 3 = 23$ is correct.

12. Divide both sides by 2: $x + 6 = 10$. Subtract 6: $x = 4$. Check: $2(4 + 6) = 2(10) = 20$ ✓.

13. Sale price is $\frac{2}{3}p$. After the \$5 coupon: $\frac{2}{3}p - 5 = 15$.

14. Subtract 6: $-3n \leq -6$. Divide by -3 and flip the sign: $n \geq 2$.

15. Add 3: $\frac{x}{2} > 4$. Multiply by 2: $x > 8$. Open circle at 8, shade right.

16. Scale ratio: $\frac{12}{4} = 3$. New width: $0.5 \times 3 = 1.5$ in.

17. The sides must add to 12. Many dimension pairs work: 1×11, 2×10, 3×9, etc. Perimeter alone does not fix the rectangle.

18. Two sides and the angle between them is the SAS (Side-Angle-Side) condition.

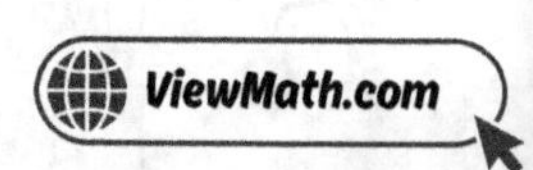

19 A vertical cut through the apex of a cone produces a triangle (specifically an isosceles triangle).

20 Full circle area $= \pi r^2 = 3.14 \times 64 = 200.96$ cm². Quarter circle $= 200.96 \div 4 = 50.24$ cm².

21 Rectangle: $6 \times 4 = 24$ cm². Triangle: $\frac{1}{2} \times 6 \times 2 = 6$ cm². Total: $24 + 6 = 30$ cm².

22 A cube has 6 faces. $SA = 6s^2 = 6 \times 16 = 96$ cm².

23 $V = Bh = 24 \times 7 = 168$ cm³.

24 The sample is the smaller group chosen from the population — the 100 students measured.

25 Class A median $= 78$, Class B median $= 65$. Class A's range (60–95) overlaps with Class B's range (50–80), but Class A is higher overall.

26 Group A range $= 20$, Group B range $= 60$. Group B's data is far more spread out.

27 There are $3 + 5 + 2 = 10$ marbles total. Blue $= 5$. $P = \frac{5}{10} = \frac{1}{2}$. Both B and C are equivalent, but $\frac{5}{10}$ directly matches the bar graph reading.

28 Predicted: $\frac{1}{4} = 0.25$. Observed: $\frac{25}{80} = 0.3125$. They are close, which supports the model.

29 Only one outcome out of 8 is all tails (TTT). $P = \frac{1}{8}$.

30 Choose 2 of the 6 faces to represent the event. $P = \frac{2}{6} = \frac{1}{3}$.

✅ Practice Test 10 — Answer Key

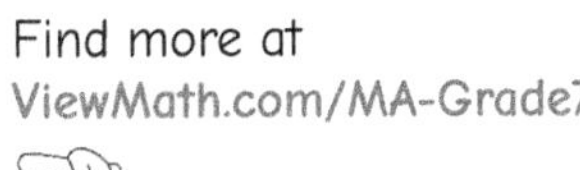
Find more at
ViewMath.com/MA-Grade7

| 1 B | 2 B | 3 C | 4 C | 5 C | 6 Day 1 with 25% error | 7 23 | 8 0 |

| 9 B | 10 B | 11 $x = 13$ | 12 B | 13 $6 | 14 C | 15 A | 16 15 cm | 17 B |

| 18 SAS; yes, the triangle is unique | 19 B | 20 B | 21 B | 22 C | 23 6 cm | 24 C |

| 25 B | 26 Median = 300; IQR = 200 | 27 D | 28 B | 29 B | 30 A |

💡 Time to Learn! 💡

Review the explanations below, **especially for the questions you missed**.

Understanding why each answer is correct builds stronger problem-solving skills.

Tip: Circle any questions you got wrong, then read their explanation carefully.

📖 Practice Test 10 — Detailed Explanations

1 Use any labeled point: $k = \frac{2}{3}$ from $(3, 2)$. Check: $\frac{4}{6} = \frac{2}{3}$ ✓.

2 The point $(1, 9)$ directly gives the unit rate: 9 miles per hour. Check: $\frac{27}{3} = 9$ and $\frac{45}{5} = 9$ ✓.

3 Set up a proportion: $\frac{2}{15} = \frac{7}{d}$. Cross-multiply: $2d = 105$, so $d = 52.5$ miles.

4 $35\% = 0.35$. Multiply: $0.35 \times 480 = 168$ fiction books.

5 Each share: $70 \div 2 = \$35$. Tip: $35 \times 0.15 = \$5.25$. Total each: $35 + 5.25 = \$40.25$.

6 Day 1: $|10 - 8|/8 = 25\%$. Day 2: $|6 - 5|/5 = 20\%$. Day 3: $|8 - 10|/10 = 20\%$. Day 1 has the largest percent error at 25%.

Find more at
ViewMath.com/MA-Grade7

7 The absolute value of -23 is its distance from 0, which is 23.

8 -18 and 18 are additive inverses (opposites). Their sum is always 0.

9 Substitute: $2(-3) + 7 = -6 + 7 = 1$.

10 Combine variable terms: $3y - 5y = -2y$. Combine constants: $8 - 2 = 6$. Result: $-2y + 6$.

11 The model shows $3x + 6 = 45$. Subtract 6: $3x = 39$. Divide by 3: $x = 13$.

12 Distribute: $8x - 6 = 26$. Add 6: $8x = 32$. Divide by 8: $x = 4$. Check: $2(4 \cdot 4 - 3) = 2(13) = 26$ ✓.

13 Student $C = \frac{1}{2} \times 12 = \$6$. Check: Student $B = 31.50 - 12 - 6 = \$13.50$. Total: $12 + 13.50 + 6 = 31.50$ ✓.

14 Subtract 9: $-x \leq 6$. Multiply by -1 and flip: $x \geq -6$.

15 Subtract 6: $-2x \leq -10$. Divide by -2 and flip: $x \geq 5$. Closed circle at 5, shade right.

16 Scale ratio: $\frac{12}{4} = 3$. New length: $5 \times 3 = 15$ cm.

17 All circles with radius 5 cm are congruent — they have the same shape and size. In terms of the figure itself (ignoring position), there is exactly one such circle.

18 Two sides (8 cm and 6 cm) and the included angle (65°) give the SAS condition. SAS always produces exactly one unique triangle.

19 A diagonal plane that passes through all six faces of a cube creates a regular hexagonal cross-section.

Find more at
ViewMath.com/MA-Grade7

ViewMath.com

20 $A = \pi r^2 = \frac{22}{7} \times 49 = 154\ in^2.$

21 Rectangle: $10 \times 4 = 40\ cm^2$. Triangle: $\frac{1}{2} \times 10 \times 3 = 15\ cm^2$. Total: $40 + 15 = 55\ cm^2$.

22 $SA = 2(30 \times 20) + 2(30 \times 6) + 2(20 \times 6) = 1200 + 360 + 240 = 1{,}800\ cm^2.$

23 $360 = 12 \times w \times 5 = 60w.$ $w = 360 \div 60 = 6\ cm.$

24 Average the two samples: $\frac{18+22}{2} = 20$ out of $50 = 40\%$. Then 40% of $1{,}000 = 400$.

25 The median is less affected by skew than the mean, so it is the best measure to compare a skewed and a symmetric distribution.

26 Median (3rd value) $= 300.$ $Q_1 = 200,\ Q_3 = 400.$ $IQR = 400 - 200 = 200.$

27 A probability of 1 means the event will definitely happen — it is certain.

28 Red covers half the spinner ($180°$), while Blue and Green each cover a quarter ($90°$ each). This is a non-uniform model.

29 Pairs with product 12: $(2,6), (6,2), (3,4), (4,3)$ — that is 4 out of 36. $P = \frac{4}{36} = \frac{1}{9}.$

30 $P(sum = 7) = \frac{6}{36} = \frac{1}{6}.$ Expected: $36 \times \frac{1}{6} = 6$. The simulation gave 8, slightly above expected.

Well done checking your answers!

Keep practicing to strengthen your skills.

Find more at
ViewMath.com/MA-Grade7

Author's Final Note

I hope you enjoyed this book as much as I enjoyed writing it. Whether you are a student working through the material, a parent supporting your child's learning, or a teacher guiding your class, I have tried to make this book as clear and engaging as possible. I hope I have succeeded. If you have any suggestions for improvement, please let me know. I would love to hear from you.

The accuracy of calculations is very important to me. We have done our best, but I also expect that I have made some minor errors. Constant improvement is the name of the game. If you find any errors, please let me know. I will fix them in the next edition.

For students: Your learning journey does not end here. I have written a series of books to help you learn math. Make sure you browse through them. I especially recommend workbooks and practice tests to help you prepare for your exams.

For parents: Thank you for investing in your child's education. I encourage you to explore the companion resources available online to help support your child outside the classroom.

For teachers: Thank you for the invaluable work you do every day. I hope this book serves as a useful resource in your classroom. Feel free to reach out if you have suggestions or would like to discuss how best to use this book with your students.

I also enjoy reading your reviews. If you have a moment, please leave a review on where you found this book. It will help others find this book. If you have any questions or comments, please feel free to contact me at drNazari@ViewMath.com.

And one last thing: Remember to use online resources for additional help. I recommend using the resources on https://ViewMath.com You can find video lessons, practice problems, and more. You can also use the online companion for this book to track your progress and access additional resources.

Wishing all students the best in their studies, parents every success in supporting their children, and teachers continued inspiration in their classrooms!

Dr. A. Nazari

 **Great Job! Keep Learning with ViewMath!**

Keep up the great work! Visit **viewmath.com/MA-Grade7** for free lessons, quizzes, and more.

Study Guide

Workbook

Step-by-Step

3 Practice Tests

5 Practice Tests

7 Practice Tests

 Find more at
ViewMath.com/MA-Grade7

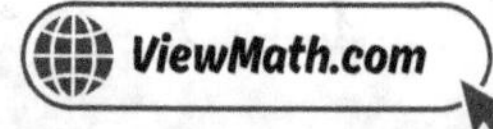 **ViewMath.com**